Brain–Machine Interface Engineering

Brain–Machine Interface Engineering

Justin C. Sanchez and Jose C. Principe

ISBN: 978-3-031-00493-3   paperback
ISBN: 978-3-031-01621-9   ebook

DOI 10.1007/978-3-031-01621-9

A Publication in the Springer series

SYNTHESIS LECTURES ON BIOMEDICAL ENGINEERING #17

Lecture #17

Series Editor: John D. Enderle, University of Connecticut

Series ISSN

ISSN 1930-0328 print

ISSN 1930-0336 electronic

Brain–Machine Interface Engineering

Justin C. Sanchez and José C. Principe
University of Florida

SYNTHESIS LECTURES ON BIOMEDICAL ENGINEERING #17

ABSTRACT

Neural interfaces are one of the most exciting emerging technologies to impact bioengineering and neuroscience because they enable an alternate communication channel linking directly the nervous system with man-made devices. This book reveals the essential engineering principles and signal processing tools for deriving control commands from bioelectric signals in large ensembles of neurons. The topics featured include analysis techniques for determining neural representation, modeling in motor systems, computing with neural spikes, and hardware implementation of neural interfaces. Beginning with an exploration of the historical developments that have led to the decoding of information from neural interfaces, this book compares the theory and performance of new neural engineering approaches for BMIs.

KEYWORDS

neural interfaces, brain, neural engineering, neuroscience, neural representation, motor systems

Foreword

"Life can only be understood backward, but it must be lived forward."
—Soren Kierkegaard

What has the past decade taught us about mankind's ability to interface with and read information from the brain? Looking back on our experiences, the salient recollection is how ill-prepared the present theories of microelectronic circuit design and signal processing are for building interfaces and interpreting brain's activity. Although there is plenty of room for future improvement, the combination of critical evaluation of current approaches and a vision of nueroengineering are helping us develop an understanding on how to read the intent of motion in brains.

The flow of ideas and discovery conveyed in this book is quite chronological, starting back in 2001 with a multi-university research project lead by Dr. Miguel Nicolelis of Duke University to develop the next-generation BMIs. The series of engineering developments explained in this book were made possible by the collaboration with Miguel, his contagious enthusiasm, vision, and brilliant experimentalism, that have led us in a journey of discovery in new theories for interfacing with the brain. Part of the results presented here also utilize data collected in his laboratory at Duke University.

It was also a journey of innovation shared with colleagues in ECE. Dr. John Harris was instrumental in designing the chips and proposing new devices and principles to improve the performance of current devices. Dr. Karl Gugel helped develop the DSP hardware and firmware to create the new generation of portable systems. We were fortunate to count with the intelligence, dedication, and hard work of many students. Dr. Justin Sanchez came on board to link his biomedical knowledge with signal processing, and his stay at University of Florida has expanded our ability to conduct research here. Dr. Sung-Phil Kim painstakingly developed and evaluated the BMI algorithms. Drs. Deniz Erdogmus and Yadu Rao helped with the theory and their insights. Scott Morrison, Shalom Darmanjian, and Greg Cieslewski developed and programmed the first portable systems for online learning of neural data. Later on, our colleagues Dr. Toshi Nishida and Dr. Rizwan Bashirullah open up the scope of the work with electrodes and wireless systems. Now, a second generation of students is leading the push forward; Yiwen Wang, Aysegul Gunduz, Jack DiGiovanna, Antonio Paiva, and Il Park are advancing the scope of the work with spike train modeling. This current research taking us to yet another unexplored direction, which is perhaps the best indication of the strong foundations of the early collaboration with Duke.

This book is only possible because of the collective effort of all these individuals. To acknowledge appropriately their contributions, each chapter will name the most important players.

Jose C. Principe and Justin C. Sanchez

Contents

CHAPTER 1

Introduction to Neural Interfaces

INTRODUCTION

Neural interfaces are one of the most exciting emerging technologies to impact biomedical research, human health, and rehabilitation. By combining engineering and neurophysiological knowledge with an innovative vision of biointeractive neuroprosthetics, a new generation of medical devices is being developed to functionally link large ensembles of neural structures in the central nervous system (CNS) directly with man-made systems. By communicating directly with the CNS, researchers are acquiring new scientific knowledge of the mechanisms of disease and innovating treatments for disabilities commonly encountered in the neurology clinic. These include spinal cord injury, stroke, Parkinson's disease, deafness, and blindness, to name a few. The seamless integration of brain and body in the healthy human makes us forget that in some diseases, normal brains can be deprived of sensory inputs (vision, audition) or even become locked in by the body (traumatic injury to the spinal cord or by degenerative diseases of the nervous system).

Several approaches have been proposed for dealing sensorimotor injuries of the nervous system, and the technologies fall into one of two categories: biological- and engineering-based solutions. In the biological category,[1] cellular and molecular techniques are being developed to regenerate and repair the damaged or dysfunctioning tissue [1–4]. Although the biological approach may ultimately provide the most natural solution to nervous system injury, the current state of research is at an early stage (i.e., spinal cord regeneration in some cases is limited in spatial scale and often operates over long time scales. Many of the new technologies shown on the right branch of Figure 1.1 have been developed to focus on the facilitation (neurotrophic factors) guidance (scaffolding) of growth of existing neurons or replacement cells (stem cells). This approach primarily works at the cellular or subcellular level of abstraction through modification of biological signaling and growth.

---

[1]The study of repair and regeneration of the central nervous system is quite broad and includes contributions from molecular/cellular neuroscience, tissue engineering, and materials science. For a comprehensive review of the application of each of these to the repair of the nervous system, see References [1–4].

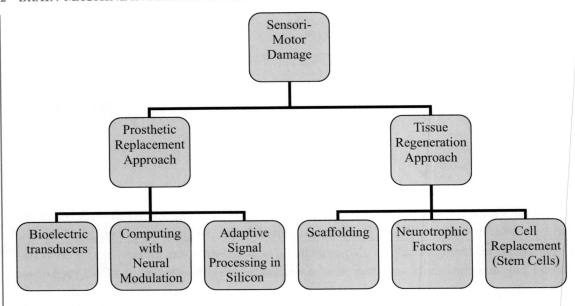

FIGURE 1.1: Roadmap of treatments for neural prosthetics, regeneration, and repair.

In contrast, the engineering approach is based on the artificial neural interface concept to provide therapy options for these individuals. While preserving the same function of the damaged biological systems, the man-made neural interface effectively creates a conduit to transduce the signaling of the nervous systems into quantities that can be used by computing architectures that augment or replace existing neural substrates [5].[2] Neural interface technologies are heavily influenced by electrical and computer engineering principles because they provide the basis and tools for the analysis of time varying quantities such as those produced by the nervous system. On the left branch of Figure 1.1, new processes must be created to mimic the integration, transduction, and interpretation of real biological events that occur internally or through interaction of the organism with the environment. These techniques include the use of adaptive signal processing to compute with neural modulation and the subsequent implementation of that processing in hardware. The overall role of the neural prosthetic are to create one or many of the following: (1) an artificial sensory system (somatosensory, occipital, auditory) by delivering the external world stimuli to the appropriate cortices; (2) an artificial command and control link that allows the motor commands to directly interact with the environment (through robotic or computerized devices); (3) a neural augmentation link within

---

[2]For a comprehensive review of sensory and motor prosthetics, see Reference [5].

or between neural networks affected by disease. In view of the amazing plasticity of the brain, these man-made devices may, with appropriate sensory feedback, be assimilated and become part of the user's cognitive space, at par with the rest of the body. Because of their man-made qualities, neural interfaces carry special qualities. Potentially, neural interfaces can scale human's natural reaction time, force, and abilities through engineered devices that are much faster and more powerful than biological tissue. In particular, they may enable a higher bandwidth between the human brain and the digital computer, which has conditioned the present research direction and metaphors for the use of computers in our society.

1.1   TYPES OF BRAIN–MACHINE INTERFACES

In the last decade, neural prosthetic technologies have commonly been grouped under the title of brain–computer interfaces (BCIs) or brain–machine interfaces (BMIs), and the names are often used interchangeably.[3] BMIs can be divided into three basic types, depending upon the application: sensory, cognitive, or motor. Sensory BMIs activate sensory systems with artificially generated signals that translate physical quantities. The most common type of sensory BMI, with more than 50,000 implanted devices [6], are cochlear implants that use miniature microphones and signal processing to translate sound wave pressure (physical quantities) in the ear into neuronal firings (neural representation) applied to the auditory nerve's tonotopic organization, allowing deaf people to listen. The same basic concept is being developed for retinal prosthesis, which can deliver to the visual system the appropriate stimulation that indicates the morphology of objects [7]. Cognitive BMIs attempt to reestablish proper neural interactions among neural systems that have damaged internal functioning within the network or do not have the ability to communicate with other networks. Cognitive BMIs have been applied for the treatment of the limbic system that may have long- and short-term memory deficits as in Alzheimer's disease [8]. Motor BMIs, on the other hand, seek to translate brain activity from the central or peripheral nervous system into useful commands to external devices. The idea is to bypass the damaged tissue (as in a spinal cord transection) and deliver the motor control commands either directly control a computer/prosthetic limb or deliver natural motor control of muscles via functional electrical stimulation [9]. Reestablishing the control of lost movement with the peripheral nervous system shares many of the neurotechnology and challenges of BMIs for prosthetic control.

Motor, cognitive, and sensory BMIs for direct neural control have far reaching impact in rehabilitation. If the whole brain or some of its parts are spared from disease or injury and the path of

---

[3]Historically, BCI has been associated with noninvasive approaches of interfacing with the nervous system as with the electroencephalogram (EEG) (described later). Alternatively, BMI commonly refers to invasive techniques.

transduction from the sensory organs (hair cells in hearing or photoreceptor cells in vision) or to the muscles (spinal cord injury or peripheral neuron damage) is broken, the following neural engineering question is appropriate: "What is the best way to deliver stimuli to the sensory cortex or to extract information about voluntary movement from motor cortex activity?" Obviously, there are two parts to the answer: one relates to the actual design and implementation of the engineering systems that can translate sound or light into electrical stimuli, control motor neurons in the periphery, or directly activate a prosthetic and the other relates to the neurophysiology knowledge to help prescribe the spatiotemporal characteristics of these stimuli to mimic the ones created by the sensory pathway to the brain or, in the case of the motor BMIs, to help model the spatiotemporal interactions of neural activity in the motor cortex. The disciplines of engineering and neuroscience (at least) are involved, and none of them independently will be able to solve the problem. Therefore, this is an imminently multidisciplinary enterprise that requires large research groups to be productive.

1.2   BEYOND STATE-OF-THE-ART TECHNOLOGY

One may think that the engineering side is better prepared to deliver systems in the short term because of the computational speed of modern digital signal processors (DSPs) and technological innovation in sensors, miniaturization, and computers, but this is illusory; interfacing with the central or peripheral nervous systems required beyond state-of-the-art technology and modeling on many fronts. Engineers working in BMI rapidly develop an appreciation for the solutions encapsulated in biological systems because of the demanding specifications of the systems. To relatively assess the relationships between the performance specifications, let us look at a space with three axes (as shown in Figure 1.2) that relates scale, power dissipation, and computation needed to interface brain tissue and microelectromechanical computer technology.

From the figure, we can see that the relationship of the space spanned by the current technology is very different from that of the BMI system, and only a small portion of the spaces overlap (purple). If we compare the scale of our current multiple-input–multiple-output (MIMO) systems, there is a large disparity because the brain is a huge computational system, with, on the order of 10^{12} neurons, about 10^{15} synapses [10]. The density of neurons in the neocortex of a human is about $10^4/mm^3$ but can vary as a function of the type of animal and the spatial location (i.e., visual, somatosensory, etc.). The conventional electrophysiological techniques (discussed later) present only a coarsely subsampling the neuronal population. The electronics needed to acquire the activity of this small sampling are very sophisticated. First, the amplifiers have to provide high common-mode rejection ratio, high gains with very small power. Second, the system has to transmit very large data rates also with constraints in power. To make the transmission more reliable, the analog signal must be converted to a digital representation, which adds substantially to the total power consumed.

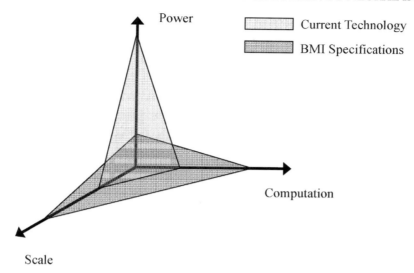

FIGURE 1.2: Comparison of system specifications for current technologies and those needed for BMIs.

Typical numbers for the transmission per channel are around ~144 kbits/channel (12.5 kHz × 12 bits). For a 16-electrode probe, this data rate is around 2.3 Mbits/sec. To compute with the electrophysiological signals at line level, front-end amplifiers (40 dB gain) used with neural probes take about 100 μW of power per channel. If one wants to transmit the data via a wireless implanted device, the total power budget can range between 2.5 and 5 mW. The area of such devices is roughly 0.5–5 mm², depending on the number of channels that is absolutely overshadowed by the biological size/power factor [11, 12]. To extend operation times of the BMI over days to weeks using hundreds of neural inputs, the power budgets need to significantly decrease. For the envisioned applications, the systems will have computational demands that exceed by orders of magnitude the capacity of a single powerful workstation. Depending on the scope of the system, it may be necessary to use hundreds or even thousands of processors to provide the needed computational power. Grid-computing infrastructures can potentially deliver these necessary resources on demand. However, the systems will also have stringent real-time requirements, as a result of the need for low latency between brain signaling and sensory feedback.

Improvements using the present approaches are very likely to be evolutionary instead of revolutionary. We have not yet discussed two fundamental and very difficult issues: how to reliably interface with the neural tissue and how to model appropriately the interactions among neurons such that proper signal processing models can be designed and applied. The neural electrode interface is of fundamental importance because invasive BMIs require prolonged contact with the neural tissue and the electrical characteristics of the interface should be stable and reliable to guarantee long life, high signal-to-noise ratios (SNRs), and avoid cell death. By far, the most compelling explanation for signal degradation in chronic neural recordings is the series of events that occurs after implantation, which include inflammatory response, a disruption of the blood–brain barrier, and initiation of astrogliosis and recruitment of microglia and macrophages to the insertion site [13–16]. Each of these problems has been studied extensively, and a variety of approaches have been proposed to improve the electrode tissue interface, which include novel electrode material coatings (conductive polymers, ceramic matrices, timed release of dexamethasone) and the use of microfluidics for engineering the glial response and tissue inflammation [5, 17]. Several groups have demonstrated that it is possible to collect data from neurons for months, but there is a long road still ahead to control tissue responses to the implant.

1.3   COMPUTATIONAL MODELING

The main problem that directly interests the readers of this book is the issue of modeling neural interactions, which will be discussed extensively throughout the text. At this point, we would like to only summarize the state of the art in computational neuroscience and a broad-stroke picture of the signals used in BMIs. There is extensive neuroanatomical knowledge on the excitability properties of neurons as well as the functional mapping of the interconnection among cortical and subcortical networks [18–20]. The issue is that the interaction principles among neurons, which implement the so-called neural code [21], are poorly understood. Moreover, large pieces of a theoretical framework that defines neural representation (i.e., mechanisms of stimulus activated elements), computation (i.e., how the elements optimally use the stimulus), and dynamics (i.e., how the elements change in time) are missing.

From a pragmatic point of view, the fundamental constructs of the brain (or any biological system for this matter) do not come with an instruction manual. The task of finding out how the system is working and what is being computed has been a significant difficulty for computational neuroscientists. Compared with most mechanical or electrical systems, the problem is perhaps even harder because of the self-organizing bottom-up principles embedded in biology versus the top-down approach used in engineering design. The difference in the two perspectives, as shown in Figure 1.3, is the necessary adoption of a big-picture approach by computer scientists and a reductionist approach by many neuroscientists. The top-down approach arises from an engineering perspective where the broad goals are to "design a machine to perform a particular task." The bottom-up perspective is one of the phenomenologist or the taxonomist: "collect data and organize

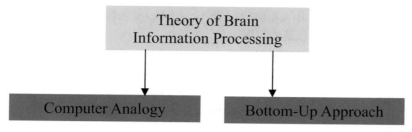

FIGURE 1.3: Bifurcation in the development of computational models of the brain.

it." In either case, the computational approaches used in the development of BMIs contain four pillars: hypothesis of how the brain functions, development of a model, analytically or numerically evaluation of the models, and testing against experimental data. In this framework, modeling plays a big part in the inductive or deductive reasoning used to establish functional associations among neural assemblies. For this reason, the models can take several forms: computational, which focus on which computation is to be performed, in terms of optimality, modularity and others, algorithmic, which look at nature of computations performed, mechanistic, that are based on known anatomy and physiology, descriptive, which summarizes large amounts of experimental data, or interpretive, which explores behavioral and cognitive significance of nervous system function.

The integrative computational neuroscience perspective and the latest technology innovation have been consistently applied to explain brain function, at least since the time of Rene Descartes. Perhaps, the formal computer metaphor of Turing and Ince [22] and Von Neumann [23] is still the one that has more followers today, but others are also relevant, such as the cybernetic modeling of Wiener [24], the universal computer analogy of McCulloch and Pitts [25], the holographic approach of Pribram [26], the connectionist modeling of McClelland and Rumelhart [27], and Rumelhart [28], Rumelhart et al. [29], and Hopfield [30], and the synergetic modeling of Haken [31] and Freeman [32], to name some important examples. This description adapted from Freeman, captures well the difficulty of the modeling task: "Unlike existing computational systems, brains and the sensory cortices embedded within them are highly unstable. They jump from one quasi-stable state to the next at frame rates of 5–10/sec under internal phase transitions. The code of each frame is a 2D spatial pattern of amplitude or phase modulation of an aperiodic (chaotic) carrier wave. The contents of each frame are selected by the input and shaped by synaptic connection weights embodying past learning. These frames are found in the visual, auditory, somatic, and olfactory cortices. Having the same code, they readily combine to form multisensory images (gestalts) that provide a landscape of basins and attractors for learned classes of stimuli."

1.4   GENERATION OF COMMUNICATION AND CONTROL SIGNALS IN THE BRAIN

The generators of the "landscape of basins and attractors," as described by Freeman, are the electro-physiological modulation of neural networks in the living tissue. From a methodological perspective, access to the spatial scale has played an important role in BMI development and includes the use of electrophysiological methods to extract spikes (action potentials) [33], local field potentials (LFPs) [34], electrocorticogram (ECoG) [35], and EEG [36].[4] Figure 1.4 shows the relative relationships of the spatial resolution for the three most common electrophysiological techniques used for BMIs. At the highest resolution, microwire electrode arrays can sense the activity of single neurons simultaneously; however, they are the most invasive to the tissue. ECoG recording is less invasive because the electrodes rest directly on the cortex, and EEG electrodes are the least invasive because they sit on the scalp. Deriving signatures of sensation or intent from electrophysiological recordings carries either the constraints of model selection or the uncertain hypotheses of the signal generation at each level of scale. For instance, microelectrode arrays that sense extracellular action potentials are very specific to neuronal function but are local in space and carry the bottleneck of sensing a representative ensemble of neurons within the technical limitations of electrode spacing and number and the difficulty of bridging the time scale to cognitive events. Because BMI paradigms have developed, the viewpoint from the single neuron has expanded to encompass the time-dependent communication of ensembles of hundreds of neurons [33]. These principles have been directly applied with great success to BMI architectures [37–40] that utilize only a single-unit activity (SUA). Recently, the use of LFPs have also been shown to improve BMI performance when combined with an SUA [34]. Therefore, harnessing both the inputs (dendritic activity) and outputs (action potentials) of neural assemblies seems to be critical for interpreting the intent of the individual.[5] Using a much larger scale, scalp EEG contains a summation of dendritic and axonal currents of millions of neurons over the cortical impedance. The continuous amplitude macroscopic signals modulate at the time scale of behavior but lack specificity and can be much more influenced by noise. The ECoG, when compared with the EEG, has the great advantage of being more spatially specific (smaller electrodes), much less contaminated by noise, and can display better frequency resolution because of its proximity to the neural tissue.

[4]Electrophysiological recordings are one particular subset of brain sensing methodologies. Other techniques such as the magnetoencephalogram, near-infrared spectroscopy, and functional magnetic resonance imaging have been used in BMI studies. However, the aforementioned techniques require quite specialized hardware and provide much different spatial and temporal resolutions compared with the electrophysiological techniques.

[5]In the next chapter, we will present a brief review of the generation of electrical potentials in excitable neural tissue.

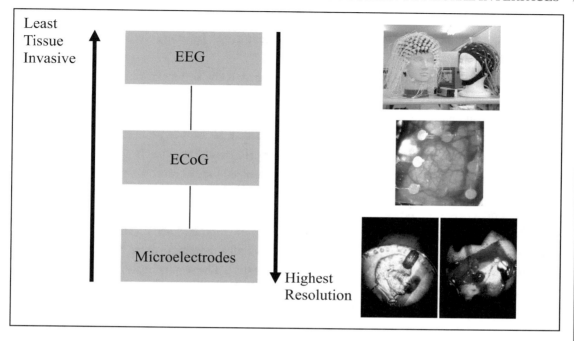

FIGURE 1.4: A comparison of the scales of electrophysiological recording methodologies.

If one analyzes the time course of one channel of data collected from the scalp (EEG), the signal time structure is very complex as shown in Figure 1.5a [41], reminding us of a nonconvergent dynamic system, or outputs of complex systems excited by bursty white noise. One of the characteristics is the rapid fluctuations observed in the signals, which, in the parlance of time series analysis, represents nonstationary phenomena and can depend upon behavioral statistics. Characterizing and extracting features from the EEG with conventional modeling is, therefore, very difficult, and the large number of publications addressing a myriad of methods well exemplifies the difficulty and suggests that the field is still seeking a good methodology to describe brain waves [42–44]. The EEG is called a macroscopic measure of brain activity, and its analysis has been shown to correlate with cognitive state [45] and brain diseases such as epilepsy [46]. The signal is created by a spatial temporal average of the pulse to amplitude conversion created when billions of pulses of current (the spike trains) are translated into voltage by the cortical tissue impedance and recorded by a 1-cm diameter electrode. The EEG frequency band extends from 0.1 to approximately 200 Hz, with a $1/f$ type of spectra (suggesting a scale free origin) and spectral peaks concentrated at a handful of frequencies where the EEG rhythms exist (delta, theta, alpha, sigma, beta, and gamma). Before being collected by the scalp electrode, the signal is attenuated and spatially filtered by the pia, dura, the corticospinal fluid, bone, and scalp. It is also contaminated by many electrical artifacts of biological or nonbiological origins.

Alternatively, if the electrodes are invasively placed at the surface of the cortex as in ECoG, the SNR is much better and the useable bandwidth increases up to at least 1 kHz (as shown in Figure 1.5c) using electrodes of 4 mm in diameter [47]. Theoretical analysis outlined by Freeman [32, 48–52] and Nunez [53–55] has identified the utility of ECoG potentials and attempts to explain how to extract the relevant modulation of neural assemblies. Closely spaced subdural electrodes have been reported to measure the spatially averaged bioelectrical activity of an area much smaller than several square centimeters [56]. The production of potentials is because of the superposition of many aligned and synchronous dipole sources [53]. The coactivation of sources is related to neural "synchrony" and is used to describe the amplitude modulations in extracellular recordings that occur during state changes [36]. The fluctuating cortical potentials have been associated with traveling waves of local dendritic and axonal potentials [57]. The results in Reference [57] indicate that at least two separate sources of signal coherence are produced either through the action of short-length axonal connections or the action of long distance connections. Synchronous activity can occur at different spatial scales and over time lags [53], which requires wide spatial sampling of the cortex and analysis over fine temporal scales.

Being a multiscale system, brain activity can also be collected at an intermediate level, which targets the neural assembly and is named the LFP as shown in Figure 1.5b. LFPs are continuous amplitude signals also in the frequency range of 0.1–200 Hz created by the same principles as the EEG, but they are normally collected with higher impedance microelectrodes with cross-sectional recording diameters of 20–100 μm. Therefore, the spatial averaging operated by the larger EEG/ECoG electrodes is not present. The LFPs differ in structure depending where the microelectrode tip is placed with respect to the dendritic or axonal processes of the neural assembly and therefore better describe the electrical field fluctuations in a more localized volume of tissue. In the frequency domain, LFP tuning has been shown to be manifested also, and differentially, in oscillatory activities in different frequency ranges. Oscillations and spiking in sensorimotor cortex in relation to motor behavior in both humans [36, 58] and nonhuman primates [59–61] in the 15- to 50-Hz range were found to increase in relation to movement preparation and decrease during movement execution [62]. Other researchers have also described the temporal structure in LFPs and spikes where negative deflections in LFPs were proposed to reflect excitatory, spike-causing inputs to neurons in the neighborhood of the electrode [63].

At the microscopic scale, the electrical activity captured by microelectrode arrays placed closer to the neuron cell body displays a pulse (called a spike) whenever the neuron fires an action potential shown in Figure 1.5b. Spikes are approximately 1 msec in duration, 30–200 μV in

(a)

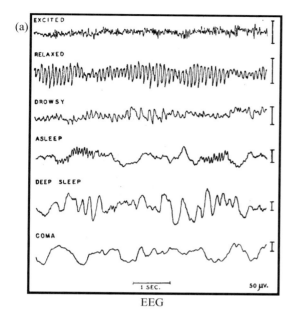

EEG

(b)

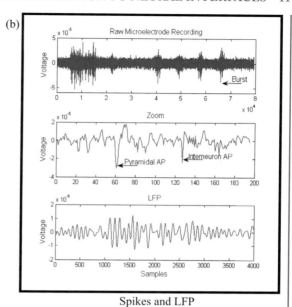

Spikes and LFP

(c)

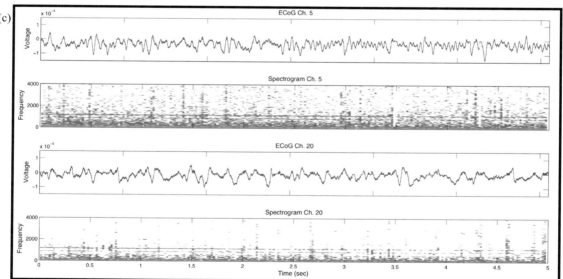

ECoG

FIGURE 1.5: Examples of electrophysiological recordings. (a) Human EEG (from Jasper and Penfield [41]), (b) spikes and LFPs from a rat (time range: top, 6.5 sec; middle, 16 msec; bottom, 328 msec), (c) ECoG from a human performing a reaching task. The patient implanted with ECoG electrodes was undergoing epilepsy monitoring.

peak-to-peak amplitude on average, and they are biphasic or triphasic in morphology depending upon the relationship of the electrode with respect to the cell body. Because of the electrophysiological morphology of the action potential (1-msec biphasic wave), the signals contain frequency in the range of 300 Hz to 7 kHz; therefore, it is common to high-pass filter the signal collected by the electrode at 300 Hz. The specificity of the spike firing is unparalleled (i.e., one can spike sort [64] the data to attribute it to a specific neuron). SUA can then be modeled as a Poisson-distributed point process [65–68]. The tradeoffs from a signal processing perspective embedded in the LFP and spike representation should be clear: LFPs loosely measure the electrical field characteristics of a region of brain tissue and the time varying activity of one or more neural populations; therefore, they are unspecific about a single neural behavior. On the other hand, the spike trains are very specific about the excitability of a given neuron, but they coarsely subsample the neural population. Unfortunately, the relationships between spike trains and LFPs are not well understood. It is reasonable to model LFP, ECoG, and EEG as nonstationary, perhaps nonlinearly generated random processes, whereas SUAs require a point process modeling.

From this discussion, it should also be apparent that there is insufficient functional knowledge of neural population dynamics in neuroscience to help guide a principled engineering approach to create realistic models of brain function or even of neural population dynamics. Therefore, we are left with the conventional engineering, statistics, and physics tools to quantify physical phenomena. Naturally, the predominance of linear approaches, Gaussian statistics, can be expected, but they should be taken with a grain of salt because they have been developed for a set of constraints likely not operational in brain function. In fact, the golden rule taken from years of working in this area is that justifying with data the assumptions of the modeling approach is critical for progress. The time dimension is of utmost importance for the brain because its role is to anticipate what is going to happen in an unpredictable and complex world. Therefore, assumptions that disregard variability over time (such as the stationarity assumption in random processes) should be avoided. Unfortunately, the tools of nonlinear dynamics are not sufficiently developed to help in this process, but we hope that someday they will come to age.

1.5 MOTOR BMIS

Because of the breadth of the BMI topic, this book will be specifically focusing on motor BMIs and will only describe properties of neurons sampled from the motor system and signal processing models and algorithms for SUA inputs. The model inputs either use point processes per se or a time-to-amplitude conversion that is called *counting* in point processes [67] or *rate coding* (binning) in neurosciences [69] that allows the use of conventional time series analysis as found in signal processing. A brief summary of BMIs for command and control using the EEG will be provided here and in the appendix for completeness.

At the center of motor BMI technology is the ability to infer the user's intent directly from the CNS with sufficient resolution and accuracy to seamlessly control an external robotic device. A conceptual drawing of a motor BMI is shown in Figure 1.6 where neural activity is recorded, conditioned, transmitted (wired or wirelessly), and translated directly into control commands of a prosthetic arm or cursor of a computer. As described earlier, variety of signal processing techniques have been used to analyze the control commands collected from EEG [70], EcOG [35], LFPs [71], and ensembles of single neurons (SUA) [72].

There are two basic types of motor BMIs: command and trajectory control BMIs. Research in command BMIs started in the 1970s with EEG [73] where subjects learned (and practice extensively) to control their regional brain activity in a predetermined fashion so that it could be robustly classified by a pattern recognition algorithm and converted into a set of discrete commands. Often, EEG-based interfaces use evoked potentials that study the impulse response of the brain as a dynamic system [74–76]. Event-related potentials are very important as clinical tools, but they are based on repeatability across stimuli, often abstract imagery, and may not be the most natural interface for real-time motor control. Therefore, the paradigm for communication is mostly based on selection of cursor actions (up/down, left/right) on a computer display. The computer presents to the users a set of possibilities, and they choose one of them through the cursor movement, until

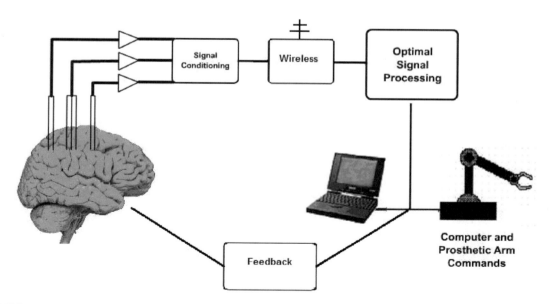

FIGURE 1.6: Motor BMIs derive intent from brain signals to command and control devices such as computers and prosthetics.

an action is completed. BCIs require subject training through biofeedback, and they display a low bandwidth for effective communication (15–25 bits/min) [77], which hinders the speed at which tasks can be accomplished. However, because of their noninvasiveness, BCIs have already been tested with success in paraplegics and locked-in patients. Several groups all over the world have demonstrated working versions of EEG-based interfaces [77], and a system software standard has been proposed. There is a biannual BCI meeting and even a BCI competition to evaluate the progress of algorithms in this area. A special issue of the *IEEE Transactions on Biomedical Engineering* also surveyed the state of the art in its June 2004 issue.

Trajectory control BMIs use brain activity to control directly the path of an external actuator in three-dimensional space, mimicking the role of the motor system when it controls, for instance, the trajectory of a hand movement. The control of trajectories is a common problem encountered by engineers (aircraft stability, gun control, etc.); therefore, feedback controllers in the area of physiological motor control are common. An example block diagram of a simplified motor system is given in Figure 1.7 [78]. The technical control problem here, at least in a first approach, does not lend itself to classify brain signals, but requires the BMI to generate a trajectory directly from brain activity. Here, the motor command generator could be the neural representation of the dynamics of the controlled object, whereas the motor command is the series of signals sent through the motor neurons, brainstem, spinal cord, and muscle fibers. In real-world situations, the controllers are subject to perturbations (loads, noise, moving objects) that can introduce errors in the original commands. To overcome these discrepancies, sensory systems can be used to modify the motor program. The two sensory modalities described here, visual and proprioceptive, operate on very different time scales. Nevertheless, because timing is critical for internal adaptation of this system, it has been difficult so far to extract from the EEG the motor command signatures with sufficient spatiotemporal resolution. Therefore, invasive techniques utilizing directly neuronal firings or LFPs have been utilized. In the last decade, the introduction of new methods for recording and analyz-

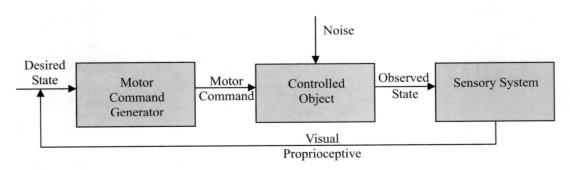

FIGURE 1.7: Elements of a feedback control system.

ing large-scale brain activity and emerging developments in microchip design, signal processing algorithms, sensors, and robotics are coalescing into a new technology—*neurotechnology*—devoted to creating BMIs for controls [33].

From an optimal signal processing viewpoint, BMI modeling is the identification of a MIMO time varying, eventually nonlinear system, which is a challenging task because of several factors: the intrinsic subsampling of the neural activity due to the relative small number of microelectrodes versus the number of neurons in the motor cortex, the unknown aspects of where the information resides (neural coding), the huge dimensionality of the problem, and the need for real-time signal processing algorithms. The problem is further complicated by a need for good generalization in nonstationary environments that depends on model topologies, fitting criteria, and training algorithms. Finally, reconstruction accuracy must be assessed, because it is linked to the choice of linear versus nonlinear and feedforward versus feedback models. In the next chapters, we will dissect the BMI design problem from a signal processing perspective where the choices in the decoding architecture will be evaluated in terms of performance and feasibility. The goal is to review and critique the past and state of the art in BMI design for guiding future development.

REFERENCES

1. Bähr, M., *Brain repair*. Advances in Experimental Medicine and Biology, 2006. **557**: pp. xii, 252.
2. Ingoglia, N.A., and M. Murray, *Axonal regeneration in the central nervous system*. Neurological Disease and Therapy, 2001. **51**: pp. xvii, 711.
3. Marwah, J., H. Teitelbaum, and K.N. Prasad, Neural Transplantation, CNS Neuronal Injury, and Regeneration: Recent Advances. 1994, Boca Raton, FL: CRC Press.
4. Seil, F.J., *Neural regeneration*. Progress in Brain Research, 1994. **103**: pp. xvi, 413.
5. Chapin, J.K., and K.A. Moxon, eds. Neural Prostheses for Restoration of Sensory and Motor Function. Methods and New Frontiers in Neuroscience. 2001, Boca Raton, FL: CRC Press.
6. Coclear, C.A., *http://www.cochlearamericas.com/About/about_index.asp*. 2004.
7. Humayun, M.S., et al., *Visual perception in a blind subject with a chronic microelectronic retinal prosthesis*. Vision Research, 2003. **43**(24): pp. 2573–2581. doi:10.1109/8.537332
8. Berger, T.W., et al., *Brain-implantable biomimetic electronics as the next era in neural prosthetics*. Proceedings of the IEEE, 2001. **89**(7): pp. 993–1012. doi:10.1109/5.939806
9. Ohnishi, K., R.F. Weir, and T.A. Kuiken, *Neural machine interfaces for controlling multifunctional powered upper-limb prostheses*. Expert Review of Medical Devices, 2007. **4**(1): pp. 43–53. doi:10.1586/17434440.4.1.43

10. Abeles, M., Corticonics: Neural Circuits of the Cerebral Cortex. 1991, New York: Cambridge University Press.

11. Akin, T., et al., *A modular micromachined high-density connector system for biomedical applications.* IEEE Transactions on Biomedical Engineering, 1999. **46**(4): pp. 471–480. doi:10.1109/10.752944

12. Wise, K.D., et al., *Wireless implantable microsystems: High-density electronic interfaces to the nervous system.* Proceedings of the IEEE, 2004. **92**(1): pp. 76–97. doi:10.1109/JPROC.2003.820544

13. Spataro, L., et al., *Dexamethasone treatment reduces astroglia responses to inserted neuroprosthetic devices in rat neocortex.* Experimental Neurology, 2005. **194**(2): p. 289. doi:10.1016/j.expneurol.2004.08.037

14. Szarowski, D.H., et al., *Brain responses to micro-machined silicon devices.* Brain Research, 2003. **983**(1–2): p. 23. doi:10.1016/S0006-8993(03)03023-3

15. Kam, L., et al., *Correlation of astroglial cell function on micro-patterned surfaces with specific geometric parameters.* Biomaterials, 1999. **20**(23–24): p. 2343. doi:10.1016/S0142-9612(99)00163-5

16. Turner, J.N., et al., *Cerebral Astrocyte Response to Micromachined Silicon Implants.* Experimental Neurology, 1999. **156**(1): p. 33. doi:10.1006/exnr.1998.6983

17. Moxon, K.A., et al., *Ceramic-Based Multisite Electrode Arrays for Chronic Single-Neuron Recording.* IEEE Transactions on Biomedical Engineering, 2004. **51**(4): pp. 647–656. doi:10.1109/TBME.2003.821037

18. Llinás, R.R., I of the Vortex: From Neurons to Self. 2000, Cambridge, MA: MIT Press.

19. Koch, C., and J.L. Davis, Large-Scale Neuronal Theories of the Brain (Computational Neuroscience). 1995, Cambridge, MA: MIT Press.

20. Buzsáki, G., Temporal Coding in the Brain (Research and Perspectives in Neurosciences). 1994, Berlin: Springer-Verlag.

21. Rieke, F., Spikes: Exploring the Neural Code. 1996, Cambridge: MIT Press.

22. Turing, A.M., and D. Ince, Mechanical Intelligence. 1992, New York: North-Holland.

23. Von Neumann, J., The Computer and the Brain. 1959, New Haven, CT: Yale University Press.

24. Wiener, N., Cybernetics; or, Control and Communication in the Animal and the Machine. 2nd ed. 1961, New York: MIT Press.

25. McCulloch, W.S., and W.H. Pitts, *A Logical Calculus for Ideas Imminent in Nervous Activity.* Bulletin of Mathematical Biophysics, 1943. **5**: pp. 115–133.

26. Pribram, K.H., Brain and Perception: Holonomy and Structure in Figural Processing. 1991, Hillsdale, NJ: Lawrence Erlbaum Associates.

27. McClelland, J.L., and D.E. Rumelhart, Explorations in Parallel Distributed Processing: A Handbook of Models, Programs, and Exercises. 1988, Cambridge, MA: MIT Press.

28. Rumelhart, D.E., Introduction to Human Information Processing. 1977, New York: Wiley.

29. Rumelhart, D.E., J.L. McClelland, and University of California San Diego PDP Research Group, Parallel Distributed Processing: Explorations in the Microstructure of Cognition (Computational Models of Cognition and Perception). 2nd vol. 1986, Cambridge, MA: MIT Press.

30. Hopfield, J.J., *Olfactory computation and object perception.* Proceedings of the National Academy of Sciences of the United States of America, 1991. **88**: pp. 6462–6466. doi:10.1073/pnas.88.15.6462

31. Haken, H., *Synergetic Computers and Cognition: A Top-Down Approach to Neural Nets.* 2nd English ed. Springer Series in Synergetics. 2004, Berlin: Springer: pp. ix, 245.

32. Freeman, W.J., Mass Action in the Nervous System: Examination of the Neurophysiological Basis of Adaptive Behavior Through EEG. 1975, New York: Academic Press.

33. Nicolelis, M.A.L., Methods for Neural Ensemble Recordings. 1999, Boca Raton, FL: CRC Press.

34. Rickert, J., et al., *Encoding of movement direction in different frequency ranges of motor cortical local field potentials.* Journal of Neuroscience, 2005. **25**(39): pp. 8815–8824. doi:10.1523/JNEUROSCI.0816-05.2005

35. Leuthardt, E.C., et al., *A brain–computer interface using electrocorticographic signals in humans.* Journal of Neural Engineering, 2004. **1**: pp. 63–71. doi:10.1088/1741-2560/1/2/001

36. Pfurtscheller, G., and F.H.L. da Silva, *Event-related EEG/MEG synchronization and desynchronization: Basic principles.* Clinical Neurophysiology, 1999. **110**(11): pp. 1842–1857. doi:10.1016/S1388-2457(99)00141-8

37. Sanchez, J.C., et al., *Ascertaining the importance of neurons to develop better brain machine interfaces.* IEEE Transactions on Biomedical Engineering, 2003. **61**(6): pp. 943–953. doi:10.1109/TBME.2004.827061

38. Serruya, M.D., et al., Brain–machine interface: Instant neural control of a movement signal. Nature, 2002. 416: pp. 141–142. doi:10.1038/416141a

39. Taylor, D.M., S.I.H. Tillery, and A.B. Schwartz, *Direct cortical control of 3D neuroprosthetic devices.* Science, 2002. **296**(5574): pp. 1829–1832. doi:10.1126/science.1070291

40. Wessberg, J., et al., *Real-time prediction of hand trajectory by ensembles of cortical neurons in primates.* Nature, 2000. **408**(6810): pp. 361–365.

41. Jasper, H., and W. Penfield, Epilepsy and the Functional Anatomy of the Human Brain. 1954, Boston: Little, Brown & Co.

42. Babiloni, F., et al., *High resolution EEG: A new model-dependent spatial deblurring method using a realistically-shaped MR-constructed subject's head model.* Electroencephalography and Clinical Neurophysiology, 1997. **102**(2): pp. 69–80. doi:10.1016/S0921-884X(96)96508-X

43. Hill, N.J., et al., *Classifying EEG and ECoG signals without subject training for fast BCI implementation: Comparison of nonparalyzed and completely paralyzed subjects.* IEEE Transactions on Neural Systems and Rehabilitation Engineering, 2006. **14**(2): pp. 183–186. doi:10.1109/TNSRE.2006.875548

44. Ginter, J., et al., *Propagation of EEG activity in the beta and gamma band during movement imagery in humans.* Methods of Information in Medicine, 2005. **44**(1): pp. 106–113.

45. Kupfermann, *Localization of higher cognitive and affective functions: The association cortices,* in: Principles of Neural Science, E.R. Kandel, J.H. Schwartz, and J.T.M, Eds. 1991, Norwalk, CT: Appleton & Lange: pp. 823–838.

46. Nakasatp, N., et al., *Comparisons of MEG, EEG, and ECoG source localization in neocortical partial epilepsy in humans.* Electroencephalography and Clinical Neurophysiology, 1994. **91**(3): pp. 171–178. doi:10.1016/0013-4694(94)90067-1

47. Sanchez, J.C., et al., *Extraction and localization of mesoscopic motor control signals for human ECoG neuroprosthetics.* Journal of Neuroscience Methods (in press). doi:10.1016/j.jneumeth.2007.04.019

48. Freeman, W.J., *Mesoscopic neurodynamics: From neuron to brain.* Journal of Physiology—Paris, 2000. **94**(5–6): pp. 303–322. doi:10.1016/S0928-4257(00)01090-1

49. Freeman, W.J., *Origin, structure, and role of background EEG activity. Part 1. Analytic Phase.* Clinical Neurophysiology, 2004. **115**: pp. 2077–2088. doi:10.1016/j.clinph.2004.02.028

50. Freeman, W.J., *Origin, structure, and role of background EEG activity. Part 2. Analytic amplitude.* Clinical Neurophysiology, 2004. **115**: pp. 2089–2107.

51. Freeman, W.J., *Origin, structure, and role of background EEG activity. Part 3. Neural frame classification.* Clinical Neurophysiology, 2005. **116**(5): pp. 1118–1129. doi:10.1016/j.clinph.2004.12.023

52. Freeman, W.J., *Origin, structure, and role of background EEG activity. Part 4. Neural frame simulation.* Clinical Neurophysiology (in press). doi:10.1016/j.clinph.2005.10.025

53. Nunez, P.L., Electric Fields of the Brain: The Neurophysics of EEG. 1981, New York: Oxford University Press.

54. Nunez, P.L., *Generation of human EEG by a combination of long and short range neocortical interactions.* Brain Topography, 1989. **1**: pp. 199–215. doi:10.1007/BF01129583

55. Nunez, P.L., Neocortical Dynamics and Human EEG Rhythms. 1995, New York: Oxford University Press.

56. Pfurtscheller, G., et al., *Spatiotemporal patterns of beta desynchronization and gamma synchronization in corticographic data during self-paced movement.* Clinical Neurophysiology, 2003. **114**(7): p. 1226. doi:10.1016/S1388-2457(03)00067-1

57. Thatcher, R.W., P.J. Krause, and M. Hrybyk, *Cortico-cortical associations and EEG coherence: a two-compartmental model.* Electroencephalography and Clinical Neurophysiology, 1986. **64**: pp. 123–143.

58. Salenius, S., and R. Hari, *Synchronous cortical oscillatory activity during motor action.* Current Opinion in Neurobiology, 2003. **13**(6): pp. 678–684. doi:10.1016/j.conb.2003.10.008

59. MacKay, W.A., and A.J. Mendonca, *Field potential oscillatory bursts in parietal cortex before and during reach.* Brain Research, 1995. **704**(2): pp. 167–174. doi:10.1016/0006-8993(95)01109-9

60. Sanes, J.N., and J.P. Donoghue, *Oscillations in local-field potentials of the primate motor cortex during voluntary movement.* Proceedings of the National Academy of Sciences of the United States of America, 1993. **90**(10): pp. 4470–4474. doi:10.1073/pnas.90.10.4470

61. Rougeul, A., et al., *Fast somato-parietal rhythms during combined focal attention and immobility in baboon and squirrel-monkey.* Electroencephalography and Clinical Neurophysiology, 1979. **46**(3): pp. 310–319.

62. Donoghue, J.P., et al., *Neural discharge and local field potential oscillations in primate motor cortex during voluntary movements.* Journal of Neurophysiology, 1998. **79**(1): pp. 159–173.

63. Arieli, A., et al., *Coherent spatiotemporal patterns of ongoing activity revealed by real-time optical imaging coupled with single-unit recording in the cat visual cortex.* Journal of Neurophysiology, 1995. **73**: pp. 2072–2093.

64. Lewicki, M.S., *A review of methods for spike sorting: The detection and classification of neural action potentials.* Network: Computation in Neural Systems, 1998. **9**(4): pp. R53–78. doi:10.1088/0954-898X/9/4/001

65. Eden, U.T., et al., *Dynamic analysis of neural encoding by point process adaptive filtering.* Neural Computation, 2004. **16**: pp. 971–998. doi:10.1162/089976604773135069

66. Brown, E.N., et al., *An analysis of neural receptive field plasticity by point process adaptive filtering.* Proceedings of the National Academy of Sciences of the United States of America, 2001. **98**(12): pp. 12 261–12 266. doi:10.1073/pnas.201409398

67. Perkel, D.H., G.L. Gerstein, and G.P. Moore, *Neuronal spike trains and stochastic point processes: I. The single spike train.* Biophysical Journal, 1967. 7(4): pp. 391–418.

68. Perkel, D.H., G.L. Gerstein, and G.P. Moore, *Neuronal spike trains and stochastic point processes: II. Simultaneous spike trains.* Biophysical Journal, 1967. **7**(4): pp. 419–440.

69. Nawrot, M., A. Aertsen, and S. Rotter, *Single-trial estimation of neuronal firing rates—From single neuron spike trains to population activity.* Journal of Neuroscience Methods, 1999. **94**: pp. 81–92. doi:10.1016/S0165-0270(99)00127-2

70. Pfurtscheller, G., and A. Aranibar, *Evaluation of event-related desynchronization (ERD) preceding and following voluntary self-paced movement.* Electroencephalography and Clinical Neurophysiology, 1979. **46**: pp. 138–146.

71. Mehring, C., et al., *Inference of hand movements from local field potentials in monkey motor cortex.* Nature Neuroscience, 2003. **6**(12): pp. 1253–1254. doi:10.1038/nn1158

72. Fetz, E.E., *Are movement parameters recognizably coded in the activity of single neurons.* Behavioral and Brain Sciences, 1992. **15**(4): pp. 679–690.

73. Vidal, J.J., *Towards direct brain–computer communication.* Annual Review of Biophysics and Bioengineering, 1973. **2**: pp. 157–180.

74. Makeig, S., *Gamma-band event-related brain dynamics—Historic perspective.* International Journal of Psychophysiology, 1993. **14**(2): pp. 136–136. doi:10.1016/0167-8760(93)90202-Z

75. Lutzenberger, W., et al., *Dimensional analysis of the human EEG and intelligence.* Neuroscience Letters, 1992. **143**(1–2): pp. 10–14. doi:10.1016/0304-3940(92)90221-R

76. Johnson, R.N., et al., *Evoked-potentials as indicators of brain dynamics—results from an interactive computer system.* Annals of Neurology, 1977. **1**(5): pp. 500–501.

77. Wolpaw, J.R., et al., *Brain–computer interfaces for communication and control.* Clinical Neurophysiology, 2002. **113**: pp. 767–791. doi:10.1016/S1388-2457(02)00057-3

78. Trappenberg, T.P., Fundamentals of Computational Neuroscience. 2002, New York: Oxford University Press.

CHAPTER 2

Foundations of Neuronal Representations

Additional Contributors: Sung-Phil Kim, Yiwen Wang

The development of BMIs relies upon the ability to decipher neuronal representations expressed by individual neuronal elements operating in distributed ensembles, and communicating using convergent and divergent principles.[1] Understanding the properties of information transmission and determining how this information translates into commands of the motor system as a whole is one of the cornerstones of BMI development. The intricate computational neurobiology of motor BMIs can be elucidated by first specifying the general principles of neural representation, transmission, and signal processing in the system, which are summarized in the following questions:

- What are the functional computational elements?
- What are the relevant mechanisms of information transmission in the system?
- How is activity aggregated between single neurons and other neurons in the system?
- How can one sufficiently model the interaction and function of biologically complex neurons and neural assemblies?

2.1 CYTOARCHITECTURE

Most of what is known about the role of neurons as the computational elements of the nervous system has been derived by measuring the cause and effect of experimental situations on neuronal firing patterns. From these initial functional observations, the neurons belonging to the neocortex that generate movement programs were spatially divided into four lobes (frontal, temporal, parietal, and occipital) and have later been associated with motor imagery [1], visual/tactile manipulation

---

[1]Groups of neurons have a variety of means of organizing their inputs. *Convergence* refers to neural input from a diverse set of sources, whereas *divergence* refers to a neuron's ability to exert influence on multiple target sites.

of objects [2], spatial coordinate transformations [3], and motor hemispheric specializations [4–6]. Historically though, the theoretical support for these functional divisions in the context of neuronal organization for computational motor control is largely incomplete. Namely, the challenge is to determine what is the neuronal substrate of mental functions [7] and how the system contributes to the principal organization of elements that are supporting mental processes such as perception and motor control. Early on, the likes of Santiago Ramón y Cajal and German anatomist Korminian Brodmann believed that the diversity in cytoarchitecture was sufficient to sustain the complexities in mental abilities. Using this rationale, Cajal studied the intricate morphology of single neuron using the Golgi method [8][2] and suggested that neurons communicate with each other via specialized synaptic junctions [9].[3] This hypothesis became the basis of the neuron doctrine, which states that the individual unit of the nervous system is the single neuron. Broadmann later also classified 52 cortical areas based upon detailed analysis of cytoarchitecture through histological analysis. However, with the advent of new electrophysiological experimental paradigms, it became clear that the structural differences did not reflect the diversity in functional specialization. In other words, the anatomically distinguishable layers of the cortex did not seem to map to all of the information-processing capabilities of neuronal networks for the following reasons. First, the columnar structure of neural assemblies is similar throughout the cortex [10]. The comparison of local processing in each of the cortices revealed similar functions: signal integration and broadcasting.

2.2 CONNECTIONISM

The apparent differences of function seem to be consequences of diverse larger scale interconnectivity and communication, which are implicitly defined in the structure. Hence, structure defines function. From a global perspective of connectivity, the main pathways among cortical and subcortical structures are primarily specified by genetics, but the fine details of the connections are influenced by the interactions of the brain with the body and environment. The fundamental architecture of the pathways consists of multiple recurrent connections of neurons within a nested hierarchy of parallel circuits. If one were to attempt to simulate the connectivity of a real neural system, to link each of the neurons in the brain to each other would require an exceeding large number of connections with impractical volume[4] and energetic requirements. The problem then becomes one of studying the minimum number of connections needed for neural networks. This problem has faced neuroscientists for many years [11], and the general solution was derived by mathematicians Erdos

---

[2] Cassic staining techniques respond to different cells or parts of cells: Nissl, color cell body; Golgi, nerve cells and fibers; Weigert, myelin, to name a few.

[3] The term *synapse* was coined by Sherrington in 1897.

[4] Axons occupy more volume in the brain than the combined contributions of cell bodies and dendrites.

and Renyi in their study of graph theory [12]. In the biological system, each neocortex neuron communicates with 5000–50,000 other neurons mostly in its proximity, but distant connections are also present, creating a random graph structure that can be modeled as a small world [13]. The density of neurons in the human neocortex is so large (10,000–100,000/mm³ [14]) that despite of the huge divergent properties, each neuron only communicates with 1% of its neighbors within its dendritic radius. Moreover, as the number of neurons in the network increases, the minimum number needed to connect them decreases.

From a functional perspective, the principal pyramidal cells (~85%) that make up these connections are primarily excitatory, and the neurons predominantly communicate by mutual excitation. However, in the absence of inhibition, the activity of the population could increase in this one direction to an unstable state. In general, excitatory activity begets excitatory activity in a positive direction. Therefore, excitatory population dynamics must be balanced by inhibitory neurons. Inhibitory activity is inherently more sophisticated than excitatory because it has several modes of operation. First, in direct negative feedback of a neuron, stability is enhanced by inhibition. Second, if the inhibitory neuron operates in a feedforward manner on another neuron, then it acts a filtering mechanism by dampening the effects of excitation. Finally, if the inhibitory neuron acts as a segregator (i.e., lateral inhibition) among neural assemblies, it can serve to function as a gating mechanism between the groups of neurons. The intricate relationships between the excitatory principal cells, inhibitory neurons, and the inherent refractory periods (i.e., when a neuron fires, it can not fire again immediately after even if fully excited) are at the root of the admirable stability of the spontaneous cortical activity.

2.3   NEURAL SIGNALING AND ELECTRIC FIELDS OF THE BRAIN

The mechanism of neural connectionism and communication involves neuronal signaling across a synapse (i.e., between its dendritic input and the axonal output) is mediated through a series of changes in cell membrane potential as shown in Figure 2.1. Molecular and voltage-gated signaling occurs at the synapses, which are chemical–electrical connectors between the axon of presynaptic neurons and one of many dendritic inputs of a given neuron. The axon of a firing presynaptic neuron transfers the transient current and subsequent action potential (spike) through a series of changes in membrane potential to its many synaptic endings. At the synapse, ions are released to the attached dendrite, locally increasing the electrical potential at the dendrite's distal portion. This voltage is slowly integrated with contributions from other dendritic branches and propagated until the cell body. When the electrical potential at the base of the dendritic tree crosses a threshold that is controlled by the cell body, the neuron fires its own action potential that is transferred through its axonal output to other postsynaptic neurons. Therefore, the closer the membrane is to the threshold,

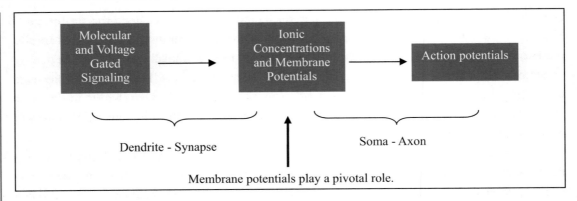

FIGURE 2.1: Relationship among neuron structure, electrochemistry, and generation of cell signaling.

the more likely is the neuron to fire. Most cortical neurons spend 99.9% of their lifetimes with their membrane potentials just below threshold.

Through this signaling procedure, there are two types of electric fields produced by neurons: waves of dendritic synaptic current that establish weak potential fields as the dendritic synaptic current flows across the fixed extracellular cortical resistance and trains of action potentials on the axons that can be treated as point processes.[5] The variation of the membrane potential initiated from a presynaptic spike can be described as a postsynaptic potential as shown in Figure 2.2. Excitatory synapses increase the membrane potential in excitatory postsynaptic potentials, which occur after the diffusion process of a neurotransmitter opens the channels of N-methyl-D-aspartate or non–N-methyl-D-aspartate receptors.[6] Inhibitory postsynaptic potentials lower the rise of the membrane potential. The electrical potentials superimpose as the sum of the individual potentials as they approach the threshold. Pulse densities are monotonically dependent on wave current densities. The range is bounded by threshold at the low end and refractoriness at the high end, with a near-linear small-signal range. This is similar to the sigmoidal curve in artificial neural networks, with a major and significant difference: the nonlinearity is asymmetric with the maximal slope skewed to the right translating the neuron's excitatory nature. As shown in chapter 1, the LFP in upper layers of neocortex (II-IV), ECoG, and EEG measure the electrical field produced by the dendritic currents (also called the post synaptic potentials) of large groups of neurons. Conventionally, microelectrode

---

[5] The review of the generation of action potentials is a brief introduction to the minimum components of the process; for a more detailed description, see Reference 15.

[6] N-methyl-D-aspartate or glutamatergic receptors are excitatory, which are typically slower than non–N-methyl-D-aspartate synapses.

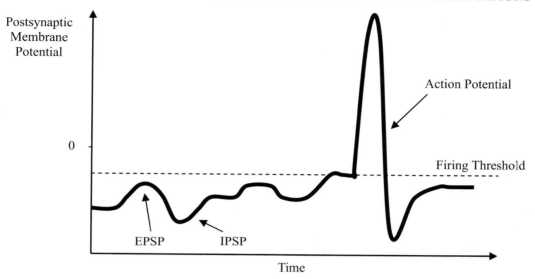

FIGURE 2.2: The time course of postsynaptic potentials. EPSP, excitatory postsynaptic potential; IPSP, inhibitory postsynaptic potential.

arrays are placed in layer V of the neocortex, close to the pyramidal/stellate cell bodies to measure their action potentials. The generation of an action potential, or spike, is considered to be the basic unit of communication among neurons. Measurement of the action potential resulting from ionic current exchanges across the membranes of neurons has been sought after for many years through the use of microelectrode recording technology [16].

An example of a microelectrode recording from a single neuron collected extracellularly near the cell body is shown in Figure 2.3a. Typical cellular potentials have magnitudes ranging from tens of microvolts to tens of millivolts, and time durations of a millisecond. The action potentials in Figure 2.3a are of the same amplitude and have the same shape, which indicates they are from the same neuron. The amplitude and shape of the measured action potential is affected by the electrode–neuron distance and orientation [17]. The ability to automatically discriminate the activity of individual neurons in a given electrode recording has been a formidable computational challenge with respect to the experimental conditional and subjective observations traditionally used to sort action potentials from each other [18]. Accurate spike detection and sorting is a critical step for high-performance BMIs because its role is to identify from the background noise the neuromodulation related to movement intent and execution. To date, accurate and automated spike detection and sorting remains an ongoing research topic [18–21]. One of the complexities of analyzing neuronal activity is the fact that neurons are active in assemblies surrounding the electrode. In Figure 2.3b,

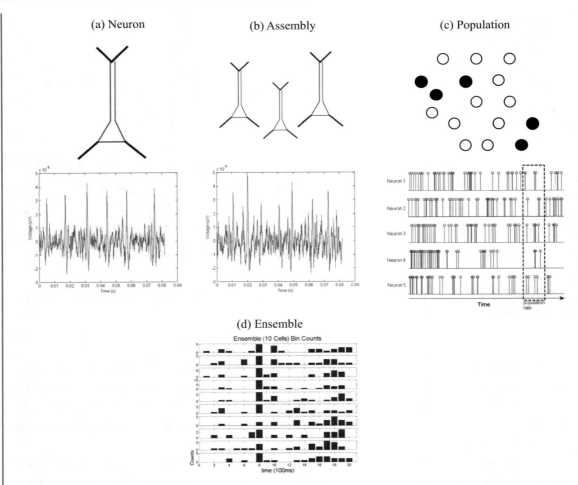

FIGURE 2.3: Levels of abstraction of neuronal representation. (a) Recordings from a single neuron. (b) A single electrode recording from multiple neurons. (c) Spike raster plots from a population of neurons indicating the time of action potentials. (d) Firing rates from an ensemble of neurons.

the action potentials of several neurons (an assembly) appear aggregated in the time series captured by one microelectrode. Here, multiple action potentials can be observed with different heights and widths. The concept of cell assembly was proposed by Hebb [52] in his book *The Organization of Behavior*, where the assembly served as a diffuse group of cells acting in a closed system that delivers facilitation to other system. In multielectrode recordings, there are many (100 are typical) channels of neural data that can translate into hundreds of neurons probed. One of the difficult aspects of interpreting this population activity is to determine information flow in networks of neurons that could be serially connected or excited through diverging or converging chains. Sherrington [9] introduced the concept of neural ensemble as a population of neurons involved in a particular computation in a distributed manner (each neuron shares a part of the processing). Because cortical neurons have large overlaps both in the extracellular potential generated by axonal projections and within dendritic trees, BMI implementations will need to interpret the many-to-one or one-to-many mappings. Moreover, the neural firing is not a deterministic process due to the tiny electrical fluctuations in the membrane produced by the other neurons, and the fact that the neuron normally operates just below threshold, waiting to fire as will be described below.

This microscopic scale of analysis can provide insight to the communication (both rate and time codes) of ensembles of local distant neurons. The variability in extracellular voltage potentials that produce variations in the neuronal recordings are attributed to cell morphology and the resistivity of the surrounding tissue. For cellular environments, the maximum measured potential has been shown to decrease in a spherical volume with a radius of 2 mm as measured from the soma of the pyramidal cell [22–24]. Experimentally and in simulation, the amplitude of the extracellular field drops off sharply at distances of 50–100 μm from the cell body [22]. Experimental access to the raw extracellular potentials enables analysis of the synaptic dendritic activity as LFPs (0.5–200 Hz) as well as spikes (300–6 kHz) originating from the cell body and axon. Depending upon the application, postprocessing of extracellular potentials is enhanced by low- or band-pass filtering. Ultimately, in single-unit neural analysis, it is the action potential that is of great interest to neural ensemble neurophysiologists. Analysis of action potentials provides a consistent waveform signature that can be tracked, which is rooted in the fundamental elements of the nervous system. Therefore, it is critical to detect peaks in the potential above the noise floor from local and distant neurons that must be assessed directly from the raw potentials. Electrophysiological parameters of the action potential (amplitude and width) are then used to identify whether a recording electrode contains any spurious signals (e.g., electrical noise, movement artifact). In a preprocessing procedure called spike sorting, the peak-to-peak amplitude, waveform shape, and interspike interval can be evaluated to ensure that the recorded neurons have a characteristic and distinct waveform shape when compared with other neuronal waveforms in the same and other recording electrodes. The goal of spike sorting is to reliably and confidently select action potentials from single neurons and extract the firing times. Hence, a variety of sorting techniques have been proposed [18], with the most common approaches being template matching and principal component analysis (PCA)–based analysis of waveform variance. Extensions of the basic techniques have been proposed using automated, manual, and multielectrode implementations [20, 25–28]. Despite the great advances in the spike sorting technique, there is often much variability in the analysis due to the experimenter [21].

2.4   SPIKING MODELS FOR THE NEURON

There are two basic types of models for the fluctuations in excitability of the neuron. The so-called conductance models that describe the channel dynamics of the ionic exchanges across the neuron membrane and the threshold-fire models[7] that try to capture the most representative behavior of neurons. One of the first conductance models was developed by Hodgkin and Huxley [29], who described the dynamics of sodium, potassium, and leak channels in the cell membrane of the giant squid axon. Building upon the contributions of Hodgkin and Huxley, there are currently many other more sophisticated compartmental neural models and many excellent software programs to simulate neural excitability [30, 31] and the dynamics of synapses [32]. One of the difficulties of this modeling approach is that it is very computational demanding because they are physics-based modeling approaches.[8] Therefore, they are mostly used to advance the study of the neuron but hardly ever used to study neural assembly activity and its relation to behavior. A notable exception is the Blue Brain Project being conducted at the Brain Science Institute at École Polytechnique Fédérale de Lausanne in Switzerland by Markram et al. who teamed with IBM to build a detailed simulation of a cortical column [33, 34].

Although the Hodgkin–Huxley model provides a detailed mathematic model on the origins of the action potential, we would like to note here that neural spikes are very stereotypical. Therefore, in the context of motor BMIs, the shape is not the information-carrying variable. Moreover, because of the computational infeasibility of modeling large networks of Hodgkin–Huxley neurons, a better modeling path may be to implicitly include the ion channel dynamics and just focus on spikes. Because organism interactions with the world are constantly changing in real time, the question now becomes one of how can discontinuous signals (spikes) be used to represent continuous motor control commands.

The threshold-fire model shown in Figure 2.4 captures the basic behavior of the neuron excitability that we are describing here (i.e., the slow integration of potential at the dendritic tree followed by a spike once the voltage crosses a threshold). In equation (2.1) $u(t)$ is the membrane potential, and $I(t)$ is the input current. In this formulation, tau serves as a membrane constant determined by the average conductances of the sodium and leakage channels. The appeal of this modeling approach is that it is computationally simple, so many neurons can be simulated in a desktop environment to simulate neural assembly behavior. The threshold-fire neuron also naturally incorporates a number of physiological parameters, including membrane capacitance, membrane

---

[7]The threshold-fire models are also known as the integrate and fire models.

[8]The physics being referred here relates to the set of differential equations that Hodgkin and Huxley derived to model each of the channel dynamics. It is the integration of these equations that adds to the computational complexity. The simulation of even 100 Hodgkin–Huxley neurons for long periods can be a formidable task.

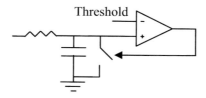

FIGURE 2.4: Threshold and fire or integrate and fire model.

(passive leak) resistance, and absolute refractory period. It is a good approximation over the normal operating range of most neurons, and it introduces the important nonlinearity (threshold) of real neurons (i.e., the neural spike). The threshold-fire neuron also has a few limitations, which include the lack of spatial extent (they are "point" neurons) and the electrochemical properties of dendrites are neglected. Membrane ionic currents tend to be nonlinear functions of both membrane potential and time. The linear summation of dendritic inputs is also inconsistent with the highly irregular firing statistics found in most cortical neurons.

$$\tau \frac{\partial u(t)}{\partial t} = -u(t) + RI(t)$$

$$(2.1)$$

This figure represents what has been called a generalized linear model, which is much more complex because of the feedback. Although the threshold-fire provides a computational approach to simulating the occurrences of spiking neurons, in practice in the electrophysiology laboratory, neuroscientists have dealt with the complexities of the raw extracellular potentials by developing a discrete representation for neuronal activity. The nature of the "all-or-nothing" firing of the action potential event has lead to the common treatment of action potentials as point processes where the continuous voltage waveform is converted into a series of timestamps indicating the instance when the spike occurred. Using the timestamps, a series of pulses or spikes (zeros or ones) can be used to visualize the activity of each neuron; this time-series (Figure 2.3c) is referred to as a *spike train*. Here, a group of electrodes are capable of capturing a subsampling of the local population (white dots) of neuronal activity indicated by the black dots.

2.5   STOCHASTIC MODELING

Single-neuron recording of this type and the analysis of spike trains reveal that most often neural responses are enormously irregular under constant stimulus. The spike trains of neural populations have several properties including sparsity and nonstationarity. This is a very different result from the threshold-fire model activity discussed previously. How do we quantify and describe the neural response variability due to the unreliable release of neurotransmitters, stochasticity[9] in channel gating, and fluctuations in the membrane potential activity? If the irregularity arises from stochastic forces, then the irregular interspike interval may reflect a random process. If we assume that the generation of each spike depends only on an underlying driving signal of instantaneous firing rate, then it follows that the generation of each spike is independent of all the other spikes; hence, we refer to this as the *independent spike hypothesis*. If the independent spike hypothesis is true, then the spike train would be completely described by a particular kind of random process called a Poisson process. Knowledge of this probability density function (PDF) would uniquely define all the statistical measures (mean, variance, etc.) of the spike train. Although the statistical properties of neural recordings can vary depending on the sample area, the animal, and the behavior paradigm, in general, spike trains modeled as Poisson distribution perform reasonably is well-constrained conditions [35]. The interspike interval from an example neuron of an animal performing a BMI task is presented in Figure 2.5a for comparison. The Poisson model has the great appeal of being defined by a single constant λ, the firing rate, assumed known

$$P_{Poisson}(x; \lambda) = \sum_{i=1}^{x} \lambda^i \frac{e^{-\lambda}}{i!} \qquad (2.2)$$

Certain features of neuronal firing, however, violate the independent spike hypothesis. Following the generation of an action potential, there is an interval of time known as the absolute refractory period during which the neuron cannot fire another spike. For a longer interval known as the relative refractory period, the likelihood of a spike being fired is much reduced. Bursting is another non-Poisson feature of neuronal spiking.

Perhaps more realistically, the firing rate λ may change over time, yielding a nonhomogeneous Poisson process [36]. One of the simplest neural models that still captures a great deal of the biological detail is the linear–nonlinear–Poisson (LNP) model shown in Figure 2.5b. This model consists of a single linear filter,[10] followed by an instantaneous nonlinear function that accounts for response nonlinearities such as rectification and saturation, followed by a Poisson generative model. The firing rate of the Poisson process is totally determined by the parameters of the linear filter and the nonlinearity. Recently, this model has been used by Simoncelli et al. [37] to estimate better the receptive field, which characterizes the computation being performed by the neuron when it responds to an input stimulus. Later, we are going to delve into more detail on this example to show

---

[9] A stochastic process is one whose behavior is nondeterministic. Specifically, the next state of the system is not fully determined by the previous state.

[10] The linear filter refers to a linear receptive field. This will be described in the next section.

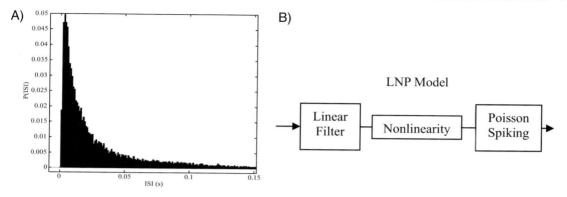

FIGURE 2.5: (a) Poisson distribution of interspike interval for an animal performing a BMI task. (b) LNP model [38].

how modern signal processing can help neuroscience data analysis methods. Indeed, it is our opinion that advanced statistical methods can enormously decrease the amount of data being collected in the wet laboratory simply by utilizing optimal statistical procedures, which are more efficient than heuristic approaches.

2.6   NEURAL CODING AND DECODING

In the previous sections, we described how information is represented in connectionist networks. For BMI motor control applications, we will now discuss how representations in populations of neurons can encode and decode stimuli. The first step is to define what are the basic constructs of neural coding? From a practical point of view, a neural code must be able to support a sufficiently rich computational repertoire. It must encode a sufficient range of values with sufficient accuracy and discriminability. Biological limitations require that the coding scheme must run under a sufficient set of operations implemented with neural mechanics. From the study of cytoarchitecture and connectionism, the representation of motor commands and sensory stimuli may be distributed or localized. The landscape of coding depends on how many components are representing the information, and three classes arise:

- Local representation: One node is active when a stimulus is presented (cardinal or grandmother cells [39]). The number of stimuli encoded increases linearly with the number of cells.
- Fully distributed representation: The stimulus is encoded by the combination of all the neurons.
- Sparsely distributed representation: Only a fraction of the neurons encode the stimuli.

Within each of these representational schemes, several operations are necessary to support a realistically complex computation of sensory information and internal representations, including the ability to combine, match, map, inverse map, compete, attenuate, and scale, to name a few. In general though, all aspects of our behavior are ultimately governed by the responses of neurons, which operate by changing their electrochemical makeup (polarizing and depolarizing). To measure how strong the relationship is between a stimulus and neural response, one can measure the correlated electrical activity between the stimulus and response. Some have framed the functional neural representation problem from a variety of perspectives, which are summarized in Table 2.1. The table is basically broken into two parts (coding and decoding), which depend upon what quantity is to be determined by the experimenter. In neural coding, a stimulus (visual and proprioceptive) is given to the system, and the neural response is to be determined. The decoding problem is the opposite, where a neural response (i.e., motor intent) is provided by the subject, and the experimenter has to determine the physical representation of the neural modulation. There are many experimental and computational approaches to elucidating the neural code that are not mutually exclusive!

The protocol that one uses to deduce the relationships in the input or output neuronal data generally falls into two classes: exploratory analysis and inference of electrophysiological recordings. Exploratory analysis is highly descriptive where graphs and numerical summaries are used to describe the overall pattern in the data. For motor BMIs, some examples of the goals of the exploratory analysis could include characterizing the spiking and LFP correlation for different behavioral conditions or explaining the spiking variation for certain behaviors in a data set. Exploratory analysis often involves characterization of a statistical distribution and can include computations such as spike rates, evoked LFP, spike-triggered averages, LFP spectral analysis, and spike-field coherence. Exploratory analysis often seeks to find patterns in nature, which can be inherently subjective. In contrast, inference is strongly hypothesis driven, where the goal is to produce neural and behavioral data in a specific way on a small scale and then draw more general conclusions about motor control for BMIs. The analysis often involves finding functional relationships in neural recordings with external quantities. Tuning curves (described later) have been used to explain the preferred kinematic spike rate for a neuron

TABLE 2.1: Intersection of experimental variables for neural coding and decoding

	Stimulus	Neural Response
Coding	Given	To be determined
Decoding	To be determined	Given

or relate LFP spectral power in a particular frequency to behavior. Regression techniques are also used to relate observed neural activity to continuous behavioral information (i.e., hand trajectory). Ultimately, hypothesis testing is required to validate inference-based approaches where significance is used to distinguish between behavioral conditions and neural activity.

2.7   METHODS OF KINEMATIC AND DYNAMIC REPRESENTATION

The firing of neurons in the primary motor (M1), premotor (PM), parietal (P), somatosensory (S) cortices, basal ganglia (BG), and cerebellum (CB) contains the representation of many motor and nonmotor signals. Through single-cell and neural population electrophysiological experiments, neuronal firing has been related to joint angles [40], muscle activation [41–43], hand trajectory (path) [44–46], velocity [47, 48], duration [49, 50], as shown in Figure 2.6. From a computational perspective, the challenge has been one of understanding how information flows in different sensory and motor networks, making contributions to each of the parameters in a time-varying manner. Several hypotheses have been proposed about how kinematic and dynamical variables are represented in the nervous system. Up to this point, we have discussed that information processing in the brain revolves around modulation (variation) in neuronal firing. This variation could be locked to some external stimulus or even the firing of another neuron. The three leading hypotheses of neuron information processing include rate coding, correlation coding, and temporal coding. A detailed discussion of the analysis approaches for each hypothesis will be given next.

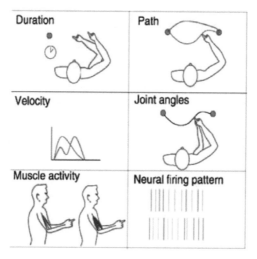

FIGURE 2.6: Parameters of cell signaling in the CNS [51].

2.7.1 Rate Coding

To understand the neuronal firing representations, neuroscientists have correlated the firing rate (2.3) with a wide variety of behaviors. The most celebrated leader of the rate coding hypothesis is English physiologist Edgar Adrian, who determined that neurons tended to modulate their firing rate in response to a stimulus (stretch receptor in the frog leg)[11] [39]. Most microelectrode-based BMIs have also utilized rate coding to map motor parameters. To represent the neural activity, most of experimental designs utilize the local mean firing rate of neurons as the input. The mean firing rate has been estimated by "binning" neural spikes with sliding rectangular time windows of length ranging from 50 up to 100 msec. This method greatly reduces the number of zeros in the spike train point process and also provides a time to amplitude conversion of the firing events. The spike rate response of cells in muscles, visual cortex, and motor cortex have lead to the development of receptive fields and tuning curves that describe the reliability of firing of a neuron as a function of the variation in an experimental condition (i.e., muscle stretch, visual orientation, movement velocity). The search for more complete mappings of receptive fields and tuning curves to the external environment has lead to the need for massively parallel neuronal recordings. These techniques yield the activity of neural ensembles or large groups of neurons from spatially disparate parts of the brain. The known concepts of the computational structure and basis of neuronal representations have been the driving force for deriving models for BMIs [14, 52–57], which will be described in the next chapter.

$$s(t) = number\ of\ spikes\ in\ \Delta t \tag{2.3}$$

2.7.2 Effect of Resolution on Rate Coding

Because the local firing rate may represent the local frequency of a neural spike train, the features can be extracted based on local frequency. One of the methods for the representation of local frequency information is multiresolution analysis [58], usually realized by the wavelet transform. Through multiresolution analysis, it is possible to represent the time-frequency characteristics of a signal. Basically, we can obtain as many local frequency components as we want at a given time instance. Hence, multiresolution analysis of neural spikes may provide richer information about neuronal behavior compared with to the binning using a fixed-width time window as is traditionally performed in the literature.

If we consider multiresolution spike train analysis, it is straightforward to see that the binning process is nothing but a discrete wavelet transform (DWT) with a Haar wavelet [59].

---

[11]Firing rate increased with increasing load on the leg.

However, because the original DWT is basically a noncausal process, a wavelet transform featuring causality should be considered. For this purpose, we develop here the à trous wavelet transform [60] to implement a causal DWT. With this procedure, multiresolution spike train analysis can be regarded as binning spike trains with multiscale windows.

2.7.3   Multiresolution Analysis of Neural Spike Trains

To facilitate the computation, we apply the DWT using dyadic Haar wavelets, which when integrated with the à trous wavelet transform, yields very effective DSP implementations. The Haar wavelet is the simplest form of wavelet and was introduced in the earliest development of wavelet transform [59]. Here, we only introduce the functional form of the Haar wavelets, whereas the details can be found in Reference [59]. Let us first define the Haar scaling function as

$$\phi(x) = \begin{cases} 1 & \text{if} \quad x \in [0,1) \\ 0 & \text{otherwise}. \end{cases} \tag{2.4}$$

Let V_j be the set of functions of the form

$$\sum_k a_k \phi(2^j x - k), \tag{2.5}$$

where a_k is a real number and k belongs to the integer set. a_k is nonzero for only a finite set of k. V_j is the set of all piecewise constant functions whose supports are finite, where discontinuities between these functions belong to a set,

$$\left\{ \cdots, -\frac{2}{2^j}, -\frac{1}{2^j}, 0, \frac{1}{2^j}, \frac{2}{2^j}, \cdots \right\}. \tag{2.6}$$

Note that $V_0 \subset V_1 \subset V_2 \subset \cdots$. The Haar wavelet function ψ is defined by

$$\psi(x) = \phi(2x) - \phi(2x - 1). \tag{2.7}$$

If we define W_j as the set of functions of the form

$$W_j = \sum_k a_k \psi(2^j x - k), \tag{2.8}$$

then it follows that

$$V_j = W_{j-1} \oplus W_{j-1} \oplus \cdots \oplus W_0 \oplus V_0 , \qquad (2.9)$$

where $\oplus$ denotes the union of two orthogonal sets.

The DWT, using a dyadic scaling, is often used because of its practical effectiveness. The output of the DWT forms a triangle to represent all resolution scales because of decimation (holding one sample out of every two) and has the advantage of reducing the computational complexity and storage. However, it is not possible to obtain representation with different scales at every instance with the decimated output. This problem can be overcome by a nondecimated DWT [61], which requires more computations and storage. The nondecimated DWT can be formed in two ways: (1) The successive resolutions are obtained by the convolution between a given signal and an incremental dilated wavelet function. (2) The successive resolutions are formed by smoothing with an incremental dilated scaling function and taking the difference between successive smoothed data.

The à trous wavelet transform follows the latter procedure to produce a multiresolution representation of the data. In this transform, successive convolutions with a discrete filter h is performed as

$$v_{j+1}(k) = \sum_{l=-\infty}^{\infty} h(l)v_j(k + 2^j l), \qquad (2.10)$$

where $v_0(k) = x(k)$ is the original discrete time series, and the difference between successive smoothed outputs is computed as

$$w_j(k) = v_{j-1}(k) - v_j(k), \qquad (2.11)$$

where w_j represents the wavelet coefficients. It is clear that the original time series $x(k)$ can be decomposed as

$$x(k) = v_S(k) + \sum_{j=1}^{S} w_j(k), \qquad (2.12)$$

where S is the number of scales. The computational complexity of this algorithm is $O(N)$ for data length N.

Note that the à trous wavelet transform as originally defined by Sheppard does not necessarily account for a causal decomposition (i.e., where the future data are not incorporated in the present computation of the wavelet transform). To apply the à trous wavelet transform with such a requirement, the Haar à trous wavelet transform (HatWT) was introduced [62]. The impulse response h is now a filter with coefficients (1/2, 1/2). For a given discrete time series, $x(k) = v_0(k)$, the first resolution is obtained by convolution $v_0(k)$ with h, such that

$$v_1(k) = \frac{1}{2}(v_0(k) + v_0(k-1)),$$ (2.13)

and the wavelet coefficients are obtained by

$$w_1(k) = v_0(k) - v_1(k).$$ (2.14)

For the jth resolution,

$$v_j(k) = \frac{1}{2}(v_{j-1}(k) + v_{j-1}(k - 2^{j-1}))$$ (2.15)

$$w_j(k) = v_{j-1}(k) - v_j(k).$$ (2.16)

Hence, the computation in this wavelet transform at time k involves only information from the previous time step $k - 1$.

The HatWT provides a set of features from the time series data, such as the wavelet coefficients $[w_1(k), \ldots, w_{S-1}(k)]$ and the last convolution output $[v_{S-1}(k)]$. However, if we seek to associate the HatWT with the binning process a spike trains, the set $[v_0(k), \ldots, v_{S-1}(k)]$ can be translated into the bin count data with multiple bin widths. To yield the multiscale bin count data using (2.13), we only have to multiply $v_j(k)$ by 2^j, such that

$$u_j(k) = 2^j v_j(k), \text{ for } j = 0, \ldots, S - 1.$$ (2.17)

Hence, the convolution output in the HatWT provides a feature set related with binning with different window widths. In the design of decoding models for BMIs, we may utilize the scaled convolution outputs $[u_j(k)]$ for $j = 0, \ldots, S - 1$, or equivalently, the bin count data with different window widths as input features.

To apply the multiresolution analysis to the BMI data, we must choose a suitable set of scales. Although it is not straightforward to determine a set of scales for the HatWT for spike trains, the firing properties of neuronal data collected from the particular BMI experimental paradigm can be used to help guide the determination of scales. For example, the smallest scale may be chosen to be larger than 1 msec because of the refractory period of neuronal firing. Also, the largest scale may be chosen

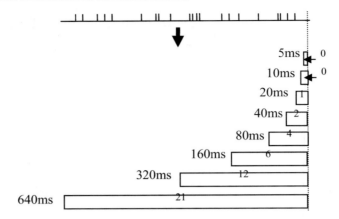

FIGURE 2.7: An illustration of the scaled convolution output from the HatWT; $u_j(k)$ for a given spike train at a time instance k_0. The number in each box denotes the value of $u_j(k_0)$ for $j = 0, \ldots, 7$.

to not exceed 1 sec because it has been reported that only the past neuronal activity up to 1 sec is correlated with the current behavior [44]. In the following case examples, we select eight scales of binning starting at 5 up to 640 msec, with the dyadic scaling of 5, 10, 20, 40, 80, 160, 320, and 640 msec.

With the selected scales, the HatWT is performed on an ensemble of BMI neuronal spike trains in the following way. We first generate the basic bin count data with a 5-msec nonoverlapping window for every neuronal channel. Next, the HatWT is applied to the 5-msec bin count data at each neuronal channel, yielding the convolution output $v_j(k)$ for $j = 0, \ldots, 7$ following (2.15). Each series $v_j(k)$ is then multiplied by 2^j to generate $u_j(k)$ [in (2.17)]. An illustrative example of the generated $u_j(k)$ at specific time instance k_0 is presented in Figure 2.7.

Note that the sampling rate in $u_j(k)$ is 200 Hz for any j. In terms of the binning process, $u_j(k)$ can be interpreted as the bin count data for a given spike train with a $5 \cdot 2^j$-msec time window that slides over time by step of 5 msec. Therefore, $u_j(k)$ with a larger j will add spikes from a longer time window and will overlap more with the previous bin $u_j(k - 1)$, which then yields a smoother temporal patterns for larger j.

The top panel in Figure 2.8 demonstrates an example of $u_j(k)$ of a specific neuron for 5-second period. $u_j(k)$ for each j is normalized to have the maximum value of 1. Darker pixels denote larger values. The set of $u_j(k)$ are temporally aligned with the associated hand trajectories plotted in the bottom panel. To view the correlation of $u_j(k)$ with the movement for each j, the discrete time series $u_j(k)$ is separately plotted on top of the hand trajectory (the x coordinate) in Figure 2.9 [both $u_j(k)$ and the hand trajectory are scaled to be in the similar dynamic range for visualization purposes].

This figure is very telling about the difficulties and opportunities of data modeling for BMIs. It demonstrates that the correlation of $u_j(k)$ with the hand trajectory increases with larger j; in fact, only the 320- and 640-msec scales have reasonable correlations with the movements. Therefore,

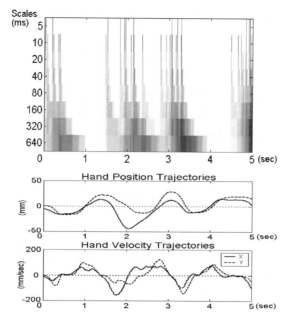

FIGURE 2.8: An example of the series of $u_j(k)$ along with the corresponding hand trajectories: (top) Each row represents the scale j for $j = 0, \ldots, 7$ nd each column represents the time indices with 5-msec interval during a 5-second duration. (bottom) The trajectories of hand position and velocity at x (solid), and y (dotted) coordinates.

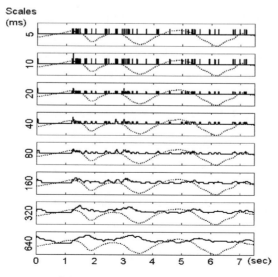

FIGURE 2.9: The demonstration of the relation between the neuronal firing activity representation at each scale (solid lines) and the hand position trajectory at x-coordinate (dotted lines).

binning at 100 msec is a reasonable value for motor BMIs when signal processing models based on correlation are used. We can expect that there are neurons that modulate their firing rates concurrently with the movement, and therefore, linear or nonlinear input output models should provide acceptable performance. But this picture also shows the difficulties between the tremendous gap in resolution between the spike trains and the movement, that are difficult to bridge with pure data driven operations (i.e., without involving some form of physiological modeling). Therefore, it opens up many other interesting possibilities to improve the present state of the art in BMI modeling.

2.7.4   Using Firing Rates to Compute Tuning

Classically, a single motor neuron's physiological response to a behavioral task has been described using directional tuning curves. As originally derived from a standard center-out task by Georgopoulos et al. [63], the tuning curve relates the mean of movement-related cell activity to movement direction. The preferred direction of a neuron, measured in degrees, is the direction that yields the maximal firing response over many trials. Tuning curves convey the expected value of a PDF indicating the average firing a cell will exhibit given a particular movement direction [63] as shown in Figure 2.10. Cells that have cosine-shaped tuning curves participate in neural computation by forming a set of basis functions[12] that represent the output. It has been shown in the literature that a variety of hand trajectories can be reconstructed by simply weighting and summing the vectors indicating preferred directions of cells in an ensemble of neurons [56, 64–66]. Weighing the preferred directions with the neural activities in the population gives a resultant direction vector called the "population vector," which has been shown to be correlated with the actual movement direction [56], and the modeling technique will be described later. To use this measurement, the kinematic data histograms must first be computed by using the desired behaviors from which the corresponding position, velocity, and acceleration vectors (magnitude and direction) between successive points in the trajectory is computed. The quantity that is commonly used in BMI experiments is the hand movement direction measured as an angle between 0° and 360° [67]. Because cellular tuning can produce properties where the average of angle 0° and 360° is not 180° (this results from the wrap-around effect of the measurements), the mean of each cell's hand direction tuning is computed using circular statistics as in (2.1)[13] [68],

$$\text{circular mean} = \arg\left(\sum_N r_N e^{i\theta_N}\right) \tag{2.1}$$

where r_N is the cell's average firing rate for angle θ_N, where N is from 1 to 360.

---

[12] A term used in vector spaces where a group of functions can be linearly combined to encode any function in a subspace.

[13] In computing the circular mean, we used the four quadrant inverse tangent.

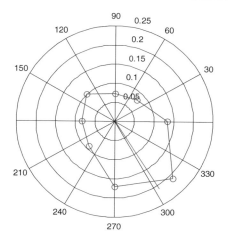
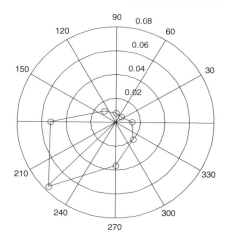

FIGURE 2.10: Examples of neuron tuning. Left plot is the tuning plot with neuronal tuning at 315° and depth of 1. The right plot is for a neuron with tuning at 225° and a depth of 0.93.

To assess the degree of tuning, a metric called the "tuning depth" is often used. This quantity is defined as the difference between the maximum and minimum values in the cellular tuning. For an impartial comparison of the cellular tuning depth, the tuning curves are usually normalized by the standard deviation of the firing rate. This measurement relates how "peaky" the tuning curve is for each cell and is an indicator of how well modulated the cell's firing rate is to the kinematic parameter of interest. The normalized tuning depth for 185 neurons computed from three kinematic vectors, hand positions, hand velocities, and hand accelerations, is presented in Figure 2.11. The cortices where the microelectrodes were implanted are also shown in the figure. From this analysis, one can clearly see that most tuned neurons are in the primary motor cortex for all of the kinematic vectors used to calculate the tuning depth.

In one-dimensional kinematic spaces, the vector space is restricted to vector magnitude. Typical histograms are ramp functions that saturate because neurons tend to increase their activity to the maximum firing rate, so the range is still finite [69]. Despite these differences, the tuning depth of each cell can still be used as a measure of information content provided by all neurons.[14] Because computing the tuning depth from finite data is noisy, the curves are also usually smoothened by

---

[14] We have observed that the tuning of a cell can vary as a function of the delay between the generation of the neural activity and the physical movement of the hand. Because we are interested in tuning depth, we are looking for cells that have the smallest variance in their tuning function. After computing tuning variance across all cells for delays up to 1 sec, we found that the sharpest tuning occurs at the 0th delay or the instantaneous time alignment.

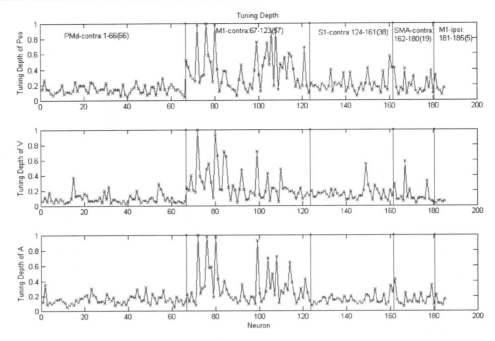

FIGURE 2.11: Tuning depth for all the neurons computed from three kinematics respectively. The upper plot is tuning depth computed from positions. The middle plot is for velocities. And the bottom plot is for accelerations.

counting the number of times that each cell fires in a window. One can expect that for BMI decoding, highly tuned cells will be more relevant. But tuning depth is not the only parameter of interest for a BMI because the input must also supply sufficient information to reconstruct all of the kinematics in the desired trajectories, that is, the set of deeply tuned neural vectors must span the space of the desired kinematic variable.

2.7.5   Information Theoretic Tuning Metrics

The tuning depth is a very coarse metric for the evaluation of the neuron receptive properties. The mutual information between neural spikes and direction angles can alternatively embody the idea of an information theoretic metric to evaluate if a neuron is tuned to a certain kinematic variable such as the angle of the velocity vector.

$$I(\text{spike}; \text{angle}) = \sum_{\text{angle}} p(\text{angle}) \sum_{\text{spike} = 0,1} p(\text{spike}|\text{angle}) \log_2 \left(\frac{p(\text{spike}|\text{angle})}{p(\text{spike})} \right) \qquad (2.2)$$

where p(angle) is the probabilistic density of all the direction angles, which can be easily estimated by the Parzen window [70]. The probability of a spike, p(spike), can be simply obtained as the percentage of the spikes occurring during a large segment of the spike train such that the estimate is reliable, and the value is not very small. p(spike | angle) is the conditional probability density of the spike given the direction angle.

For each neuron, the histogram of the spike-triggered angle can be drawn and normalized to approximate p(spike = 1, angle). In other words, only when there is a spike, the direction angle is counted for the histogram during the corresponding direction angle bin. Then p(spike = 1| angle) is approximated by dividing the histogram of (spike = 1, angle) by the histogram of (angle), which corresponds actually to the Bayesian formulation.

$$p(\text{spike} = 1 | \text{angle}) = \frac{p(\text{spike} = 1, \text{angle})}{p(\text{angle})}$$

(2.3)

and p (spike = 0|angle) = $1 - p$(spike = 1|angle) .

In this way, the mutual information between the kinematic direction angle and the neural spike train can be estimated for each neuron. Figure 2.12 shows the mutual information tuning calculated from three different kinematic vectors for 185 neurons. The M1 (primary motor cortex) tuning clearly shows that the velocity (the middle plot) conveys more tuning information than position or acceleration, which has been suggested in the literature, but it is impossible to evaluate in Figure 2.11 because the conventional method normalizes to the same value all the responses. Therefore, because of the self-normalized definition, the information theoretical tuning metric found that particular neurons are more tuned to the position, whereas other neurons are more tuned to the velocity. Because the mutual information is a nonlinear measure, experience has shown that the information theoretic tuning depth should be plotted in logarithmic scale for BMIs.

2.7.6   Timing Codes

Finding appropriate methodologies to analyze in time the spatial neuronal organization during motor control, auditory/visual processing, and cognition is a challenging question. Methods for identifying the fine timing relationships among co-activated neurons have focused on statistical relationships among observed single spike trains [71–76] (see Reference [77] for a review). For pairs of neuronal recordings, several techniques have been proposed such as cross-intensity functions [78] or the method of moments [79], joint peristimulus time histograms [80], and methods based on maximum likelihood (ML) estimation [81].

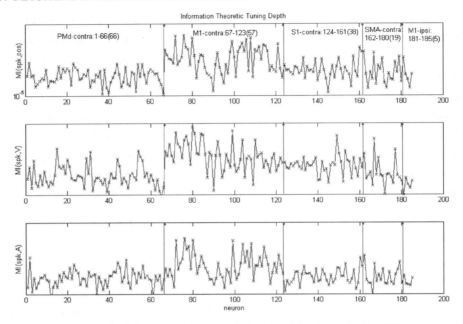

FIGURE 2.12: Information theoretic tuning depth in logarithmic scale for 185 neurons computed from three kinematic variables respectively. The upper plot is tuning depth computed from position, the middle plot is for velocity, and the bottom plot is for acceleration.

However, in many applications such as the design of neural interfaces, the timing of large ensembles of cells (hundreds of cells) must be studied in a principled way. Unfortunately, many of the aforementioned techniques become extremely time-consuming (as all pairs of electrode combinations must be considered) and alternative pattern recognition [74] or clustering [82] methods to quantify multichannel prespecified spike timing relations (patterns) have been proposed.

Most statistical analysis of neuronal spike timing assumes Poisson spike generation where the neural response only depends upon the short-term past of the input stimulus due to the independence assumptions embedded in the Poisson model. To estimate the response of a given neuron to these stimuli, one needs simply to quantify the likelihood of a spike firing in each region of the stimulus space, which can be estimated by the ratio of multidimensional histograms of spike triggered stimuli. To accurately estimate the histograms, large amounts of data are required to isolate the analysis to regions of the space where the neuron shows a large response (many counts to evaluate the histograms) or to tailor the stimuli, parameterized by only a few parameters to produce large neural responses. To overcome data issues, one can time lock the averaging to the spike, a procedure known as spike-triggered averaging. Once the spike-triggered average is obtained, the neural response can be inferred with a nonlinear input output neural model.

In conjunction with these techniques, a second class of nonlinear models for BMIs was derived from system theory using the concept of Volterra series expansions in Hilbert spaces. Volterra series have been proposed to model neuronal receptive fields using the reverse correlation or white noise approach [83], and more recently they have been applied to cognitive BMIs to model the CA3 region of the hippocampus [84]. A Volterra series is similar to a Taylor series expansion in functional spaces, where the terms are convolution integrals of products of impulse responses called kernels. The Volterra models are universal approximators, but they require the estimation of many parameters. Therefore, when applied in practice, one of the design difficulties is how to minimize the parameters, either by limiting the expansion to the second or third term or by using polynomial approximations, such as Laguerre polynomials [85] or, preferably, Kautz functions, to decrease the number of free parameters for the impulse responses.

However, because the neural response tends to be nonlinear, and higher order terms of the Volterra expansion have many parameters, lots of experimental data are needed to estimate the parameters well. The LNP (as shown in Figure 2.5) can reduce the collection of large data sets by using a more efficient estimation of the linear filter term, followed by the instantaneous nonlinearity. The assumptions for the method to work are mild: basically, one assumes that the stimuli components are uncorrelated and they are Gaussian distributed with the same variance (i.e., that the D-dimensional distribution is spherically symmetric).

In the LNP model, the linear filter collapses the multidimensional input space to a line. Spike-triggered averaging provides an estimate of this axis, under the assumption that the raw stimulus distribution is spherically symmetric [86]. Once the linear filter has been estimated, one may compute its response and then examine the relationship between the histograms of the raw and spike-triggered ensembles within this one-dimensional space to estimate the nonlinearity that now can be done with histograms without requiring large data sets. The model can be generalized to take into consideration several directions at the same time by using the covariance. Such approaches have been proposed by Moran and Schwartz introduced an exponential velocity and direction tuned motor cortical model [67]. Eden et al. used a Gaussian tuning function for the hippocampal pyramidal neurons [87]. These nonlinear mathematical models can encounter difficulties when dealing with the real data because the neurons can have very different tuning properties and probably change over time. The accuracy of the tuning function estimation can directly affect the prior knowledge of the Bayesian approach and, therefore, the results of the kinematic estimation. Simoncelli et al. built upon the approach of Marmarelis and proposed a statistical method to model the neural responses with stochastic stimuli [85]. Through parametric model identification, the nonlinear property between the neural spikes and the stimuli was directly estimated from data, which is more reliable. For motor BMIs, one is seeking sequential state estimation of the point process algorithm to infer the kinematic vectors from the neural spike train, which is the opposite of sensory neurons. However, one can regard the proper kinematic vector as the outcome of the motor cortex neurons. The tuning function between the kinematic vector and neural spike train is exactly the system model between the state and observation in our algorithm. Therefore, these methods can provide a very practical way to acquire the prior knowledge (the tuning function) in sequential estimation for adaptive point process filtering algorithms.

2.8 MODELING AND ASSUMPTIONS

2.8.1 Correlation Codes

The coincidence structure among spike trains can be used to suggest the functional connectivity among neurons (not necessarily the anatomic connectivity) in a minimal equivalent circuit [16]. Moreover, correlation and synchrony among neurons can greatly influence postsynaptic neurons, help define cell assemblies, and play a role in information transmission. Often, BMI studies seek to quantify the conditions of full synchrony and total independence among the recorded neurons. As a control, the cross correlation is often used to determine the similarities and differences in spatio-temporal organization in the data using the zero-lag cross-correlation over time, averaged through all pairwise combinations of neuron types. The cross-correlation function [88] is probably the most widely used technique to measure similarity between spike trains. With access to the activity of large neuronal ensembles one can begin to simultaneously describe the spatial and temporal varying nature of the data by computing local neuronal correlations. In this scheme, information contained in cell assemblies can be observed in cortical areas that mutually excite each other [52]. To extract useful information in the local bursting activity of cells in the ensemble, local correlations among the cells is analyzed. Traditionally, the correlogram is a basic tool to analyze the temporal structure of signals. However, applying the correlogram to spike trains is nontrivial because they are point processes, thus the signals information is not in the amplitude but only on the time instances when spikes occur. A well-known algorithm for estimating the correlogram from point processes involves histogram construction with time interval bins [88]. The binning process is effectively transforming the uncertainty in time to amplitude variability. However, this time quantization introduces binning error and leads to coarse time resolution. Furthermore, the correlogram does not take advantage of the higher temporal resolution of the spike times provided by current recording methods. However, one of the disadvantages of continuous time estimation is its computational cost. The joint peristimulus time histogram [89] requires trial averaging and, therefore, is only applicable to experimental paradigms where trial repetition is performed, and it assumes stationarity between trials. Similar to the cross-correlation, the partial directed coherence (PDC) [90–93] and the methods by Hurtado et al. [94] and Samonds and Bonds [95] suffer from the problem of windowing and large variations of the covariance matrix of the stochastic noise, which might lead to wrong conclusions about the underlying interdependence structure of the data. Typically, these methods are applied to time binned (or continuous) signals; hence, they constrain the time resolution of synchrony. Moreover, the extension of the analysis of the methods for more than two spike trains is not a trivial problem. Finally, the algorithms are difficult to apply in online applications or are computationally complex for large ensembles of neurons. The gravity transform [96] successfully tackles some of the above problems, but it lacks a statistical basis. Most importantly, none of the mentioned measures is appropriate to analyze transient synchronization because a measure in the time or spike domain is needed. As one can conclude from the discussion, still more work needs to be done to create a useful similarity metric for spike trains.

The cross-correlation function in (2.4) can be used to quantify synchronized firing among cells in the ensemble. In this case, small sliding (overlapping) windows of data are defined by $\mathbf{A}(t)$, which is a matrix-containing L-delayed versions of firing patterns from 185 neurons. At each time tick t in the simulation, the cross-correlation between neurons i and j for all delays l is computed. Because we are interested in picking only the most strongly synchronized bursting neurons in the local window, we simply average over the delays in (2.5). We define $\overline{C}_{\cdot j}$ as a vector representing how well correlated the activity of neuron j is with the rest of the neuronal ensemble. Next, a single 185×1 vector at each time tick t is obtained in (2. 6) by summing cellular assembly cross-correlations only within each sampled cortical area, M1 (primary motor), PMd (premotor dorsal), SMA (supplementary motor associative), and S1 (somatosensory).

$$C(t)_{ijl} = E\left[\mathbf{A}(t)_i\,\mathbf{A}(t-l)_j\right]$$

(2.4)

$$\overline{C}(t)_{ij} = \frac{1}{L}\sum_{l=1}^{L}\left|C(t)_{ijl}\right|$$

(2.5)

$$\tilde{C}(t)_j = \sum_{i=\text{cortex}} \overline{C}(t)_{ij}$$

(2.6)

The neural correlation measure in (2.6) was then used as a real-time marker of neural activity corresponding to segments of movements of a hand trajectory. In Figure 2.13, highly correlated neuronal activity is shown to be time-varying in synchrony with changes in movement direction as indicated by the vertical dashed line. Figure 2.13 shows that the correlations contained in the data are highly variable even for similar movements.

We can immediately see that different neurons present different level of involvement in the task as measured by correlation. Some participate heavily (hot colors), whereas others basically do not correlate with the hand motion (blue). But even for the ones that participate heavily, their role across time is not constant as depicted by the vertical structure of the raster. This is stating that the time structure of neural firings is nonstationary and heavily dependent upon the task.

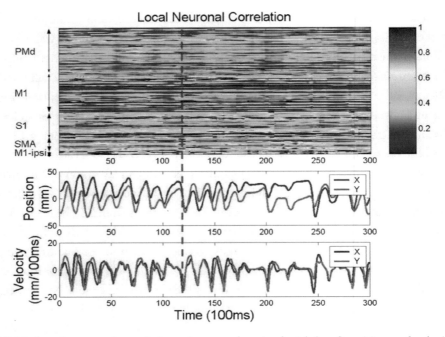

FIGURE 2.13: Local neuronal correlations time-synchronized with hand position and velocity.

2.9   IMPLICATIONS FOR BMI SIGNAL PROCESSING

One of the most important steps in implementing BMI optimal signal processing techniques for any application is data analysis and the understanding of neural representations. Here, the reader should take note that optimality in the signal processing technique is predicated on the matching between the statistics of the data and the a priori assumptions inherent in any signal processing technique [97]. In the case of BMIs, the statistical properties of the neural recordings and the analysis of neural ensemble data are not fully understood. Hence, this lack of information means that the neural-motor translation is not guaranteed to be the best possible, even if optimal signal processing is utilized (because the criterion for optimality may not match the data properties). Despite this reality, through the development of new neuronal data analysis techniques we can improve the match between neural recordings and BMI design [16, 35]. For this reason, it is important for the reader to be familiar with the characteristics of neural recordings that would be encountered.

Some interesting issues to consider include the following:

- How much information is encoded in the neural population response? Noise and the finite number of neurons sampled can influence how one interprets the representation.

- What is the most efficient decoding strategy (i.e., the one that has the minimum decoding error)?
- What are the relationships between resolution and synchronization in the neural representation? We have shown a variety of techniques that can alter the interpretation of the neural representation and depend upon time resolution incorporated in the neural recordings and the synchronization techniques with behavioral or neural events.
- What is the biological plausibility of the encoding or decoding strategy that is assumed? The constraints used in biological systems are very different than those used in computational environments. One must identify and discuss these limitations within the constructs of the BMI computational goals.

In this chapter, we have observed that the neuronal and ensemble responses are highly nonstationary (i.e., local correlations over time) and sparse (spike trains). Moreover, firing rate analysis and its correlation with behavior highly depend on the appropriate choice of the rate window. The important message of this analysis is that standard optimal signal processing techniques (linear filters, neural networks, etc.) were not designed for data that is nonstationary and discrete valued. Ideally, we would like our optimal signal processing techniques to capture the changes observed in Figure 2.13. However, the reader should be aware that in the environment of BMI datasets, applying any of the "out-of-the-box" signal processing techniques means that the neural to motor mapping may not be optimal. More importantly, any performance evaluations and model interpretations drawn by the experimenter can be directly linked and biased by the mismatch between data and model type.

REFERENCES

1. Crammond, D.J., *Motor imagery: never in your wildest dream.* Trends in Neurosciences, 1997. **20**(2): pp. 54–57. doi:10.1016/S0166-2236(96)30019-2
2. Kupfermann, *Localization of Higher Cognitive and Affective Functions: The Association Cortices*, in Principles of Neural Science, E.R. Kandel, J.H. Schwartz, and T.M. Jessel, eds. 1991, Norwalk, CT: Appleton & Lange. pp. 823–838.
3. Andersen, R.A., et al., *Multimodal representation of space in the posterior parietal cortex and its use in planning movements.* Annual Review of Neurosciences 1997. **20**: pp. 303–330. doi:10.1146/annurev.neuro.20.1.303
4. Chen, R., L.G. Cohen, and M. Hallett, *Role of the ipsilateral motor cortex in voluntary movement.* Can. J. Neurol. Sci., 1997. **24**: pp. 284–291.
5. Cisek, P., D.J. Crammond, and J.F. Kalaska, *Neural activity in primary motor and dorsal premotor cortex In reaching tasks with the contralateral versus ipsilateral arm.* Journal of Neurophysiology, 2003. **89**(2): pp. 922–942. doi:10.1152/jn.00607.2002

6. Tanji, J., K. Okano, and K.C. Sato, *Neuronal activity in cortical motor areas related to ipsilateral, contralateral, and bilateral digit movements of the monkey.* Journal of Neurophysiology, 1988. **60**(1): pp. 325–343.

7. Edelman, G.M., V.B. Mountcastle, and Neurosciences Research Program. The Mindful Brain: Cortical Organization and the Group-Selective Theory of Higher Brain Function. 1978, Cambridge, MA: MIT Press.

8. Polak, J.M., and S. Van Noorden, Introduction to Immunocytochemistry. 3rd ed. 2003, Oxford: BIOS Scientific Publishers.

9. Sherrington, C.S., The Integrative Action of the Nervous System. Classics in Psychology. 1973, New York: Arno Press. doi:10.1097/00005053-192306000-00038

10. Mountcastle, V., *The columnar organization of the neocortex.* Brain, 1997. **120**: pp. 702–722. doi:10.1093/brain/120.4.701

11. Buzsáki, G., Rhythms of the Brain. 2006, New York: Oxford University Press. doi:10.1126/science.1066880

12. Erdos, P. and R. A., *On the evolution of random graphs.* Publication of the Mathematical Institute of the Hungarian Academy of Science, 1960. **5**: pp. 17–61.

13. Watts, D. and S. Strogatz, *Collective dynamics of "smal-world" networks.* Nature, 1998. **393**: pp. 440–442.

14. Abeles, M., Corticonics: Neural Circuits of the Cerebral Cortex. 1991, New York: Cambridge University Press.

15. Kandel, E.R., J.H. Schwartz, and T.M. Jessell, eds. Principles of Neural Science. 4th ed. 2000, New York: McGraw-Hill.

16. Nicolelis, M.A.L., Methods for Neural Ensemble Recordings. 1999, Boca Raton, FL: CRC Press.

17. Holt, G.R., and C. Koch, *Electrical Interactions via the Extracellular Potential Near Cell Bodies.* Journal of Computational Neuroscience, 1999. **6**: pp. 169–184.

18. Lewicki, M.S., *A review of methods for spike sorting: The detection and classification of neural action potentials.* Network: Computation in Neural Systems, 1998. **9**(4): pp. R53–78. doi:10.1088/0954-898X/9/4/001

19. Fee, M.S., P.P. Mitra, and D. Kleinfeld, *Variability of extracellular spike waveforms of cortical neurons.* Journal of Neurophysiology, 1996. **76**(6): pp. 3823–3833.

20. Fee, M.S., P.P. Mitra, and D. Kleinfeld, *Automatic sorting of multiple-unit neuronal signals in the presence of anisotropic and non-Gaussian variability.* Journal of Neuroscience Methods, 1996. **69**: pp. 175–188.

21. Wood, F., et al., *On the variability of manual spike sorting.* IEEE Transactions on Biomedical Engineering, 2004. **51**(6): pp. 912–918. doi:10.1109/TBME.2004.826677

22. Koch, C. and G.R. Holt, *Electrical interactions via the extracellular potential near cell bodies.* Journal of Computational Neuroscience, 1999. **6**: pp. 169–184.

23. Somogyvari, Z., et al., *Model-based source localization of extracellular action potentials.* Journal of Neuroscience Methods, 2005. **147**: pp. 126–137. doi:10.1016/j.jneumeth.2005.04.002

24. Varona, P., et al., *Macroscopic and subcellular factors shaping population spikes.* Journal of Neurophysiology, 2000. **83**(4): pp. 2192–2208.

25. Bierer, S.M., and D.J. Anderson, *Multi-channel spike detection and sorting using an array processing technique.* Neurocomputing, 1999(26–27): pp. 946–956. doi:10.1016/S0925-2312(99) 00090-9

26. Takahashi, S., Y. Anzai, and Y. Sakurai, *A new approach to spike sorting for multi-neuronal activities recorded with a tetrode—How ICA can be practical.* 2003. **46**(3): p. 265. doi:10.1016/S0168-0102(03)00103-2

27. Rogers, C.L., et al. *A Pulse-Based Feature Extractor for Spike Sorting Neural Signals*, in 3rd International IEEE EMBS Conference on Neural Engineering. 2007. Kohala Coast, HI.

28. Rao, S., et al. *Spike Sorting Using Non Parametric Clustering Via Cauchy Schwartz PDF Divergence*, in ICASSP. 2006.

29. Hodgkin, A. and A. Huxley, *A quantitative description of membrane current and its application to conduction and excitation in nerve.* Journal of Physiology, 1952. **117**: pp. 500–544.

30. Koch, C., Biophysics of computation; information processing in single neurons. 1999, New York: Oxford University Press.

31. Koch, C. and I. Segev, eds. Methods in Neural Modelling. 1998, New York: MIT Press.

32. Wilson, H.R., Spikes Decision and Actions: Dynamical Foundations of Neuroscience. 1999, New York: Oxford University Press.

33. Grillner, S., et al., *Microcircuits in action—From CPGs to neocortex.* Trends in Neurosciences, 2005. **28**(10): pp. 525–533. doi:10.1016/j.tins.2005.08.003

34. *Blue Brain Project.* 2007 (available from: http://bluebrain.epfl.ch/).

35. Rieke, F., Spikes: Exploring the Neural Code. 1996, Cambridge, MA: MIT Press.

36. Brown, E.N., et al., *The time-rescaling theorem and its application to neural spike train data analysis.* Neural Computation, 2002. **14**(2): pp. 325–346. doi:10.1162/08997660252741149

37. Simoncelli, E.P., et al., *Characterization of neural responses with stochastic stimuli.* 3rd ed. The New Cognitive Neuroscience. 2004, Cambridge, MA: MIT Press.

38. Pillow, J.W., et al., *Prediction and decoding of retinal ganglion cell responses with a probabilistic spiking model.* Journal of Neuroscience, 2005. **25**(47): pp. 11003–11013. doi:10.1523/JNEUROSCI. 3305-05.2005

39. Adrian, E., Nobel Lectures, Physiology or Medicine 1922–1944. 1965, Amsterdam: Elsevier.

40. Evarts, E.V., *Representation of movements and muscles by pyramidal tract neurons of the precentral motor cortex*, in Neurophysiological Basis of Normal and Abnormal Motor Activities, M.D. Yahr and D.P. Purpura, eds. 1967, New York: Raven Press. pp. 215–253.

41. Todorov, E., *Direct cortical control of muscle activation in voluntary arm movements: A model.* Nature Neuroscience, 2000. **3**(4): pp. 391–398.

42. Fetz, E.E., and D.V. Finocchio, *Correlations between activity of motor cortex cells and arm muscles during operantly conditioned response patterns.* Experimental Brain Research, 1975. **23**(3): pp. 217–240. doi:10.1007/BF00239736

43. Sanes, J.N., S. Suner, and J.P. Donoghue, *Dynamic organization of primary motor cortex output to target muscles in adult rats. I. Long-term patterns of reorganization following motor or mixed peripheral nerve lesions.* Experimental Brain Research, 1990. **79**: pp. 479–491. doi:10.1007/BF00229318

44. Wessberg, J., et al., *Real-time prediction of hand trajectory by ensembles of cortical neurons in primates.* Nature, 2000. **408**(6810): pp. 361–365.

45. Sanchez, J.C., et al. *Learning the contributions of the motor, premotor, and posterior parietal cortices for hand trajectory reconstruction in a brain machine interface*, in IEEE EMBS Neural Engineering Conference. 2003. Capri, Italy. doi:10.1109/CNE.2003.1196755

46. Lin, S., J. Si, and A.B. Schwartz, *Self-organization of firing activities in monkey's motor cortex: Trajectory computation from spike signals.* Neural Computation, 1997. **9**: pp. 607–621. doi:10.1162/neco.1997.9.3.607

47. Carmena, J.M., et al., *Learning to control a brain–machine interface for reaching and grasping by primates.* PLoS Biology, 2003. **1**: pp. 1–16. doi:10.1371/journal.pbio.0000042

48. Sanchez, J.C., et al. *Simultaneous prediction of five kinematic variables for a brain-machine interface using a single recurrent neural network*, in International Conference of Engineering in Medicine and Biology Society. 2004.

49. Scott, S.H., *Role of motor cortex in coordinating multi-joint movements: Is it time for a new paradigm?* Canadian Journal of Physiology and Pharmacology, 2000. **78**: pp. 923–933. doi:10.1139/cjpp-78-11-923

50. Kalaska, J.F., et al., *Cortical control of reaching movements.* Current Opinion in Neurobiology, 1997. 7: pp. 849–859. doi:10.1016/S0959-4388(97)80146-8

51. Shadmehr, R. and S.P. Wise, The Computational Neurobiology of Reaching and Pointing: A Foundation for Motor Learning. 2005, Cambridge, MA: MIT Press.

52. Hebb, D.O., The Organization of Behavior: A Neuropsychological Theory. 1949, New York: Wiley.

53. Lilly, J.C., *Correlations between neurophysiological activity in the cortex and short-term behavior in monkey*, in Biological and Biochemical Bases of Behavior, H.F. Harlow and C.N. Woolsey, eds. 1958, Madison, WI: University Wisconsin Press. pp. 83–100.

54. Strumwasser, F., *Long-Term recording from single neurons in brain of unrestrained mammals.* Science, 1958. **127**(3296): pp. 469–470.

55. Grossberg, S., *Studies of mind and brain: Neural principles of learning, perception, development, cognition, and motor control.* Boston Studies in the Philosophy of Science. Vol. 70. 1982, Dordrecht, The Netherlands: Boston. pp. 223–225.

56. Georgopoulos, A.P., A.B. Schwartz, and R.E. Kettner, *Neuronal population coding of movement direction.* Science, 1986. **233**(4771): pp. 1416–1419.

57. Buzsaki, G., *Large-scale recording of neuronal ensembles.* Nature Neuroscience, 2004. **75**(5): pp. 446–451. doi:10.1038/nn1233

58. Murtagh, F., J.L. Starck, and O. Renaud, *On neuro-wavelet modeling.* Decision Support System Journal, 2004. **37**: pp. 475–484. doi:10.1016/S0167-9236(03)00092-7

59. Daubechies, I. *Ten lectures on wavelets*, in Society for Industrial and Applied Mathematics. 1992. Philadelphia, PA: SIAM. doi:10.1121/1.406784

60. Shensa, M.J., *Discrete wavelet transforms: Wedding the à trous and Mallat algorithm.* IEEE Transactions on Signal Processing, 1992. **40**: pp. 2464–2482. doi:10.1109/78.157290

61. Aussem, A., J. Campbell, and F. Murtagh, *Wavelet-based feature extraction and decomposition strategies for financial forecasting.* Journal of Computational Intelligence in Finance, 1998. **6**: pp. 5–12.

62. Zheng, G., et al., *Multiscale transforms for filtering financial data streams.* Journal of Computational Intelligence in Finance, 1999. 7: pp. 18–35.

63. Georgopoulos, A., et al., *On the relations between the direction of two-dimensional arm movements and cell discharge in primate motor cortex.* Journal of Neuroscience, 1982. **2**: pp. 1527–1537.

64. Georgopoulos, A.P., et al., *Mental rotation of the neuronal population vector.* Science, 1989. **243**(4888): pp. 234–236.

65. Schwartz, A.B., D.M. Taylor, and S.I.H. Tillery, *Extraction algorithms for cortical control of arm prosthetics.* Current Opinion in Neurobiology, 2001. **11**(6): pp. 701–708. doi:10.1016/S0959-4388(01)00272-0

66. Taylor, D.M., S.I.H. Tillery, and A.B. Schwartz, *Direct cortical control of 3D neuroprosthetic devices.* Science, 2002. **296**(5574): pp. 1829–1832. doi:10.1126/science.1070291

67. Moran, D.W., and A.B. Schwartz, *Motor cortical representation of speed and direction during reaching.* Journal of Neurophysiology, 1999. **82**(5): pp. 2676–2692.

68. Jammalamadaka, S.R., and A. SenGupta, Topics in Circular Statistics. 1999, River Edge, NJ: World Scientific Publishing Company.

69. Fu, Q.G., et al., *Temporal encoding of movement kinematics in the discharge of primate primary motor and premotor neurons.* Journal of Neurophysiology, 1995. **73**(2): pp. 836–854.

70. Parzen, E., *On the estimation of a probability density function and the mode.* Annals on Mathematical Statistics, 1962. **33**(2): pp. 1065–1076.

71. Perkel, D.H., G.L. Gerstein, and G.P. Moore, *Neuronal spike trains and stochastic point processes. II. Simultaneous spike trains.* Biophysical Journal, 1967. **7**(4): pp. 419–40.

72. Gerstein, G.L., and D.H. Perkel, *Simultaneously recorded trains of action potentials: Analysis and functional interpretation.* Science, 1969. **164**(881): pp. 828–30.

73. Gerstein, G.L., D.H. Perkel, and K.N. Subramanian, *Identification of functionally related neural assemblies.* Brain Research, 1978. **140**(1): pp. 43–62. doi:10.1016/0006-8993(78)90237-8

74. Abeles, M., and G.L. Gerstein, *Detecting spatiotemporal firing patterns among simultaneously recorded single neurons.* Journal of Neurophysiology, 1988. **60**(3): pp. 909–24.

75. Palm, G., A.M. Aertsen, and G.L. Gerstein, *On the significance of correlations among neuronal spike trains.* Biological Cybernetics, 1988. **59**(1): pp. 1–11. doi:10.1007/BF00336885

76. Grun, S., M. Diesmann, and A. Aertsen, *Unitary events in multiple single-neuron spiking activity: I. Detection and significance.* Neural Computation, 2002. **14**(1): pp. 43–80. doi:10.1162/089976602753284455

77. Gerstein, G.L., and K.L. Kirkland, *Neural assemblies: Technical issues, analysis, and modeling.* Neural Networks, 2001. **14**(6–7): pp. 589–98. doi:10.1016/S0893-6080(01)00042-9

78. Cox, D.R., and P.A.W. Lewis, *Multivariate point processes.* Proceedings of the Sixth Berkeley Symposium on Probability and Mathematical Statistics, 1972. **3**: pp. 401–448.

79. Brillinger, D.R., *The identification of point process systems.* Annals of Probability, 1975. **3**: pp. 909–929.

80. Gerstein, G.L., and D.H. Perkel, *Mutual temporal relationships among neuronal spike trains. Statistical techniques for display and analysis.* Biophysical Journal, 1972. **12**(5): pp. 453–473.

81. Borisyuk, G.N., et al., *A new statistical method for identifying interconnections between neuronal network elements.* Biological Cybernetics, 1985. **52**(5): pp. 301–306. doi:10.1007/BF00355752

82. Gerstein, G.L., and A.M. Aertsen, *Representation of cooperative firing activity among simultaneously recorded neurons.* Journal of Neurophysiology, 1985. **54**(6): pp. 1513–1528.

83. Marmarelis, P.Z., and V.Z. Marmarelis, Analysis of Physiological Systems: The White Noise Approach. 1978, New York, Plenum Press.

84. Song, D., V.Z. Marmarelis, and T.W. Berger. *Parametric and non-parametric models of short-term plasticity*, in Second Joint EMBS/BMES Conference. 2002. Houston, TX. doi:10.1109/IEMBS.2002.1053117

85. Marmarelis, V.Z., *Identification of nonlinear biological systems using Laguerre expansions of kernels.* Annals of Biomedical Engineering, 1993. **21**: pp. 573–589. doi:10.1007/BF02368639

86. Chichilnisky, E.J., *A simple white noise analysis of neuronal light responses.* Network: Computation in Neural Systems, 2001. **12**: pp. 199–213.

87. Eden, U.T., et al., *Dynamic Analysis of Neural Encoding by Point Process Adaptive Filtering.* Neural Computation, 2004. **16**: pp. 971–998. doi:10.1162/089976604773135069

88. Dayan, P., and L. Abbott, Theoretical Neuroscience: Computational and Mathematical Modeling of Neural Systems. 2001, Cambridge, MA: MIT Press.

89. Aertsen, A., et al., *Dynamics of neuronal firing correlation: Modulation of "effective connectivity".* Journal of Neurophysiology, 1989. **61**(5): pp. 900–917.

90. Baccala, L.A., and K. Sameshima, *Partial directed coherence: A new concept in neural structure determination.* Biological Cybernetics, 2001. **84**(6): pp. 463–474. doi:10.1007/PL00007990

91. Baccala, L.A., and K. Sameshima, *Overcoming the limitations of correlation analysis for many simultaneously processed neural structures.* Brain Research, 2001. **130**: pp. 33–47.

92. Sameshima, K., and L.A. Baccala, *Using partial directed coherence to describe neuronal ensemble interactions.* Journal of Neuroscience Methods, 1999. **94**: pp. 93–103. doi:10.1016/S0165-0270(99)00128-4

93. Baccala, L.A., and K. Sameshima, *Directed coherence: A tool for exploring functional interactions among brain structures,* in Methods for Neural Ensemble Recordings, M. Nicolelis, ed. 1999, Boca Raton, FL: CRC Press. pp. 179–192.

94. Hurtado, J.M., L.L. Rubchinsky, and K.A. Sigvardt, *Statistical method for detection of phase-locking episodes in neural oscillations.* Journal of Neurophysiology, 2004. **91**(4): pp. 1883–1898. doi:10.1152/jn.00853.2003

95. Samonds, J.M., and A.B. Bonds, *Real-time visualization of neural synchrony for identifying coordinated cell assemblies.* Journal of Neuroscience Methods, 2004. **139**(1): pp. 51–60. doi:10.1016/j.jneumeth.2004.04.035

96. Gerstein, G.L., and A. Aertsen, *Representation of cooperative firing activity among simultaneously recorded neurons.* Journal of Neurophysiology, 1985. **54**(6): pp. 1513–1528.

97. Haykin, S., Adaptive filter theory. 3rd ed. 1996, Upper Saddle River, NJ: Prentice-Hall International.

• • • •

CHAPTER 3

Input–Output BMI Models

Additional Contributors: Sung-Phil Kim, Yadunandana Rao, and Deniz Erdogmus

The BMI experimental paradigm lends itself to statistical signal processing methodologies used to derive optimal models from data. Indeed, in the BMI setting, the researcher has available synchronously both the input to the BMI (neuronal activity – spike trains) as well as the desired response (kinematics of movement). The problem can be then framed in terms of "decoding," by which spike occurrences of individual neurons are translated into intended movement. In terms of modeling, the aim is to find a functional relationship between neural activity and the kinematic variables of position, velocity, acceleration, and force. Here, we will describe how to translate the BMI decoding problem into a system identification framework, where a parametric linear or nonlinear system can be trained directly from the collected data to achieve outputs close to the kinematics as in Figure 3.1. Model building has been extensively studied in control theory and signal processing so there are a wealth of methods that can be utilized [1].

Because the data are collected by multielectrode arrays, and each electrode can potentially sense several neurons, spike sorting is commonly utilized to identify individual neurons. Provided that the spike features have been accurately extracted, the neuronal spike firings become the decoding model input. Because the desired responses are continuous amplitude signals with unknown structure, we will treat them as random processes. As for the neural data, this chapter will deal with spike rate coding in short windows of 100 msec (called binned data), and a stochastic model for the data will also be assumed. This problem can therefore be framed as system identification with stochastic processes, which is a very well developed area in optimal signal processing and controls [2]. However, its application to BMI is not without problems for the following reasons:

1. There is a huge difference in the time scale of spike trains (msec) and that of behavior (sec) that the model has to bridge. Binning the spikes firing of each neuron over an appropriate time window (50 ~ 100 msec) has been widely used to smooth the spike trains and provide a time scale closer to behavior. It also implements a simple method for time to amplitude conversion of the neural modulation.

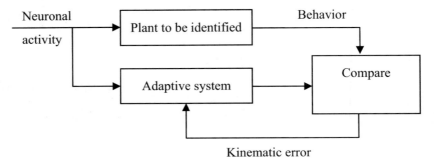

FIGURE 3.1: System identification framework for BMIs.

2. The system is MIMO, with a very large input (hundred of neurons), and the data are limited; therefore, there is a need for proper regularization of the solution.
3. Although the input dimensionality is high, it still coarsely samples in unknown ways the large pool of neurons involved in the motor task (missing data).
4. Not all the channels are relevant for the task, and because of the spike sorting and other difficulties, the data are noisy.
5. Because of the way that neurons are involved in behavior, the data are nonstationary.

With these remarks in mind, one has to choose the class of functions and model topologies that best match the data while being sufficiently powerful to create a mapping from neuronal activity to a variety of behaviors. There are several candidate models available, and based on the amount of neurophysiological information that is utilized about the system, an appropriate modeling approach shall be chosen. Three types of I/O models based on the amount of prior knowledge exist in the literature [3]:

- "White box": The physical system principles of operation are perfectly known, except for the parameters that are learned from the data.
- "Gray box": Some physical insight is available but model is not totally specified, and other parameters need to be determined from the data.
- "Black box": Little or no physical insight is available, so the chosen model is picked on other grounds (robustness, easy implementation, etc).

The choice of white, gray, or black box is dependent upon our ability to access and measure signals at various levels of the motor system, the type of questions one is interested to answer, as well as the computational cost of implementing the model in current computing hardware.

The first modeling approach, the "white box," would require the highest level of physiological detail. Starting with behavior and tracing back, the system comprises muscles, peripheral nerves, the spinal cord, and ultimately the brain. This would be a prohibitively difficult task with the present level of knowledge because of the complexity, interconnectivity, and dimensionality of the involved neural structures. Model implementation would require the parameterization of a complete motor system [4] that includes the cortex, cerebellum, basal ganglia, thalamus, corticospinal tracts, and motor units. Because all of the details of each component/subcomponent of the described motor system remain largely unknown and are the subject of study by many neurophysiological research groups around the world, it is not presently feasible to implement white-box modeling for BMIs. Even if it was possible to parameterize the system to some high level of detail, the task of implementing the system in our state-of-the-art computers and DSPs would be an extremely demanding task. The appeal of this approach when feasible is that a physical model of the overall system could be built, and many physiological questions could be formulated and answered.

The "gray box" model requires a reduced level of physical insight. In the "gray box" approach, one could take a particularly important feature of the motor nervous system or of one of its components, incorporate this knowledge into the model, and then use data to determine the rest of the unknown parameters. One of the most common examples in the motor BMI literature is Georgopoulos' population vector algorithm (PVA) [5]. Using observations that cortical neuronal firing rates were dependent on the direction of arm movement, a model was formulated to incorporate the weighted sum of the neuronal firing rates. The weights of the model are then determined from the neural and behavioral recordings.

The last model under consideration is the "black box." In this case, it is assumed that no physical insight is available to accomplish the modeling task. This is perhaps the oldest type of modeling where only assumptions about the data are made, and the questions to be answered are limited to external performance (i.e., how well the system is able to predict future samples, etc.). Foundations of this type of time series modeling were laid by Wiener for applications in antiaircraft gun control during World War II [22]. Although military gun control applications may not seem to have a natural connection to BMIs, Wiener and many other researchers provided the tools for building models that correlate unspecified time series inputs (in our case, neuronal firing rates) and desired targets (hand/arm movements). We will treat black box modeling in this chapter.

The three input–output modeling abstractions have gained a large support from the scientific community and are also a well-established methodology in control theory for system identification and time series analysis [2]. Here, we will concentrate on the last two types that have been applied by engineers for many years to a wide variety of applications and have proven that the methods produce viable phenomenological descriptions when properly applied [23, 24]. One of the advantages of the techniques is that they quickly find, with relatively simple algorithms, optimal mappings (in the sense of minimum error power) between different time series using a nonparametric approach (i.e., without requiring a specific model and only mild assumptions for the time series generation). These advantages have to be counter weighted by the abstract (nonstructural) level of the modeling and the many difficulties of the method, such as determining what is a reasonable fit, a model order, and a topology to appropriately represent the relationships among the input and desired response time series.

3.1   MULTIVARIATE LINEAR MODELS

The first black box model that we will discuss is the linear model, which has been the workhorse for time series analysis, system identification, and controls [2]. Perhaps, the first account of a filter-based BMI [25] used a linear model trained by least squares with one second (i.e., the current and NINE past bins) of previous spike data per neuron as a memory buffer to predict a lever press in a rat model. It turns out that least squares with past inputs can be shown equivalent to the Wiener solution for a finite impulse response filter. Therefore, the first BMI was trained with a Wiener filter. The bulk of the results reported in the literature use this very simple but powerful linear solution [8, 10, 26, 27]. Since for our BMI application the filter takes a slightly different form because of the multidimensional input and output, it will be presented more carefully than the others. Let us assume that an M-dimensional multiple time series is generated by a stationary stable vector autoregressive (VAR) model

$$x(n)=b+\mathbf{W}_1 x(n-1)+...+\mathbf{W}_L x(n-L)+u(n) \qquad (3.1)$$

where b is a vector of intercept terms $b = [b_1,...,b_M]^T$ and the matrices $\mathbf{W}_i$ are coefficient matrices of size $M \times M$ and $u(n)$ is white noise with nonsingular covariance matrix $\mathbf{S}$. We further assume that we observe the time series during a period of T samples. The goal of the modeling is to determine the model coefficients from the data. Multivariate least square estimation can be applied to this problem without any further assumption. Let us develop a vector notation to solve the optimization problem.

$$X=[x_1,...x_T] \qquad MxT$$
$$A=[b,W_1,...,W_L] \qquad Mx(ML+1)$$
$$Z_n=[1,x_n,...,x_{n-L+1}]^T \qquad (ML+1)x1$$
$$Z=[Z_0,...,Z_{T-1}] \qquad (ML+1)xT$$
$$U=[u_1,...,u_T] \qquad MxT$$
$$\chi=\text{vec}[X] \qquad MTx1$$
$$\alpha=\text{vec}[A] \qquad (M^2L+M)x1$$
$$v=\text{vec}[U] \qquad MTx1 \qquad\qquad (3.2)$$

where vec is a vector stacking operator that creates a vector from a matrix, one column at a time. Note that the Z vector is composed of the previous values of the input that are in the filter taps. Using this notation, the VAR (L) model can be written as

$$X = AZ + U \tag{3.3}$$

Thus, the multivariate least square estimation chooses the estimator that minimizes

$$J(\alpha) = v^{\mathrm{T}}\Sigma^{-1}v = tr\{(X - AZ)^{\mathrm{T}}\Sigma^{-1}(X - AZ)\} \tag{3.4}$$

To find the minimum of this function, we have to take the derivative with respect to the unknown parameters α

$$\frac{\partial J(\alpha)}{\partial \alpha} = 2(\mathbf{ZZ}^{\mathrm{T}} \otimes \Sigma^{-1})\alpha - 2(\mathbf{Z} \otimes \Sigma^{-1})\chi \tag{3.5}$$

where $\otimes$ is the Kronecker product of matrices [28] (which creates a matrix whose elements are given by the elements of the first matrix multiplied by the second matrix). The normal equations are obtained equating the derivative to zero, and the optimal solution becomes

$$\hat{\alpha} = ((ZZ^{\mathrm{T}})^{-1}Z \otimes \mathbf{I})\chi \tag{3.6}$$

where $\mathbf{I}$ is the identity matrix of size M. Equation (3.6), written back into matrix form, reads

$$A = X\mathbf{Z}^{\mathrm{T}}(\mathbf{ZZ}^{\mathrm{T}})^{-1}, \tag{3.7}$$

which has the same form as the conventional least square solution of the univariate time series case. Therefore, we can conclude that the multivariate LS estimation is equivalent to the univariate least square solution of each one of the M equations in (3.7), which means that the conventional adaptive filtering algorithms [2] that use least squares can be used by simply renaming the quantities involved. One can establish the asymptotic normality of the vector LS estimator provided the data matrix $\mathbf{ZZ}^{\mathrm{T}}$ is nonsingular, as well as the Gaussianity of the errors in the limit of infinite long data windows provided the noise is the standard vector white noise (i.e., mean zero, nonsingular covariance, independent, and with finite kurtosis). It is therefore possible to use the conventional t-statistics to check confidence intervals for model coefficients (the number of degrees of freedom can be estimated by T-ML-1). Furthermore, it can be shown that the vector LS estimator coincides with the maximum likelihood estimator for Gaussian distributed stochastic processes [29].

Unfortunately, the small sample behavior of the multivariate least square estimator is a lot harder to establish, and normally, Monte Carlo methods are used to establish empiric coefficient variances, and the decision about the adequacy of the fitting is left to the experimenter. In the practical situation of BMIs, more important issues require our attention. In fact, the spike train time series are not even stationary, and the mappings to the desired response are probably not linear. The approaches presented thus far assume that the spike train statistics do not change over time. This is an unrealistic assumption that can be counteracted using sample by sample adaptation [the famous least mean square (LMS) algorithm] to track the optimal solution through time as will be shown in 3.1.2 or by partitioning the solution space into a set of local linear models that switch automatically among themselves [30]. Moreover, the MIMO models have many parameters, and so generalization is a major concern. Therefore, one must proceed with great caution.

3.1.1   Linear Modeling for BMIs and the Wiener Filter

Because the VAR decomposes into a parallel set of independent linear models (one for each output), we will present here this more familiar notation. Moreover, because this solution coincides with the Wiener–Hopf solution applied to discrete time data and finite impulse responsive filters (FIRs), it will be called here the digital Wiener filter [2]. Consider a set of spike counts from M neurons, and a hand position vector $\mathbf{d} \in \Re^C$ (C is the output dimension, $C = 2$ or 3). The spike count of each neuron is embedded by an L-tap discrete time-delay line. Then, the input vector for a linear model at a given time instance n is composed as $\mathbf{x}(n) = [x_1(n), x_1(n-1) \ldots x_1(n-L+1), x_2(n) \ldots x_M(n-L+1)]^T$, $\mathbf{x} \in \Re^{L \cdot M}$, where $x_i(n-j)$ denotes the spike count of neuron i at a time instance $n-j$. A linear model estimating hand position at time instance n from the embedded spike counts can be described as

$$y^c = \sum_{i=0}^{L-1} \sum_{j=1}^{M} x_i(n-j) w_{ij}^c + b^c \qquad (3.8)$$

where y^c is the c coordinate of the estimated hand position by the model, w_{ij}^c is a weight on the connection from $x_i(n-j)$ to y^c, and b^c is a bias for the c coordinate. The bias can be removed from the model when we normalize $\mathbf{x}$ and $\mathbf{d}$ such that $E[\mathbf{x}] = \mathbf{0}$, $\mathbf{0} \in \Re^{L \cdot M}$, and $E[\mathbf{d}] = \mathbf{0}$, $\mathbf{0} \in \Re^C$, where $E[\cdot]$ denotes the mean operator. Note that this model can be regarded as a combination of three separate linear models estimating each coordinate of hand position from identical inputs. In a matrix form, we can rewrite (3.8) as

$$\mathbf{y} = \mathbf{W}^T \mathbf{x} \qquad (3.9)$$

where $\mathbf{y}$ is a C-dimensional output vector, and $\mathbf{W}$ is a weight matrix of dimension $(L \cdot M+1) \times C$. Each column of $\mathbf{W}$ consists of $[w_{10}^c, w_{11}^c, w_{12}^c \ldots, w_{1L-1}^c, w_{20}^c, w_{21}^c \ldots, w_{M0}^c, \ldots, w_{ML-1}^c]^T$.

Figure 3.2 shows the topology of the MIMO linear model for the BMI application, which will be kept basically unchanged for any linear model through this book. The most significant dif-

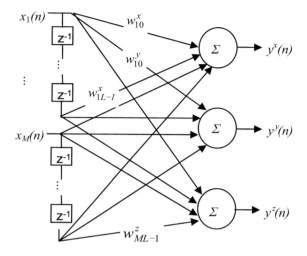

FIGURE 3.2: The topology of the linear filter designed for BMIs in the case of the 3D reaching task. $x_i(n)$ are the bin counts input from the ith neuron (total M neurons) at time instance n, and z^{-1} denotes a discrete time delay operator. $y^c(n)$ is the hand position in the c coordinate. w^c_{ij} is a weight on $x_i(n - j)$ for $y_c(n)$, and L is the number of taps.

ferences will be in the number of parameters, and in the way, the parameters w_{ij} of the model are computed from the data.

For the MIMO case, the weight matrix in the Wiener filter system is estimated by

$$\mathbf{W}_{\text{Wiener}} = \mathbf{R}^{-1}\mathbf{P} \qquad (3.10)$$

$\mathbf{R}$ is the correlation matrix of neural spike inputs with the dimension of $(L \cdot M) \times (L \cdot M)$,

$$\mathbf{R} = \begin{bmatrix} \mathbf{r}_{11} & \mathbf{r}_{12} & \cdots & \mathbf{r}_{1M} \\ \mathbf{r}_{21} & \mathbf{r}_{22} & \cdots & \mathbf{r}_{2M} \\ \vdots & \vdots & \ddots & \vdots \\ \mathbf{r}_{M1} & \mathbf{r}_{M2} & \cdots & \mathbf{r}_{MM} \end{bmatrix}, \qquad (3.11)$$

where r_{ij} is the $L \times L$ cross-correlation matrix between neurons i and j ($i \neq j$), and r_{ii} is the $L \times L$ auto-correlation matrix of neuron i. $\mathbf{P}$ is the $(L \cdot M) \times C$ cross-correlation matrix between the neuronal bin count and hand position as

$$\mathbf{P} = \begin{bmatrix} \mathbf{p}_{11} & \cdots & \mathbf{p}_{1C} \\ \mathbf{p}_{21} & \cdots & \mathbf{p}_{2C} \\ \vdots & \ddots & \vdots \\ \mathbf{p}_{M1} & \cdots & \mathbf{p}_{MC} \end{bmatrix},$$

(3.12)

where $\mathbf{p}_{ic}$ is the cross-correlation vector between neuron i and the c coordinate of hand position. The estimated weights $\mathbf{W}_{\text{Wiener}}$ are optimal based on the assumption that the error is drawn from white Gaussian distribution, and the data are stationary. The predictor $\mathbf{W}^{\text{T}}_{\text{Wiener}}\mathbf{x}$ minimizes the mean square error (MSE) cost function,

$$J = E[\|\mathbf{e}\|^2], \mathbf{e} = \mathbf{d} - \mathbf{y},$$

(3.13)

Each sub-block matrix r_{ij} can be further decomposed as

$$\mathbf{r}_{ij} = \begin{bmatrix} r_{ij}(0) & r_{ij}(1) & \cdots & r_{ij}(L-1) \\ r_{ij}(-1) & r_{ij}(0) & \cdots & r_{ij}(L-2) \\ \vdots & & \ddots & \vdots \\ r_{ij}(1-L) & r(2-L) & \cdots & r(0) \end{bmatrix},$$

(3.14)

where $r_{ij}(\tau)$ represents the correlation between neurons i and j with time lag τ. These correlations, which are the second-order moments of discrete time random processes $x_i(m)$ and $x_j(k)$, are the functions of the time difference $(m - k)$ based on the assumption of wide sense stationarity (m and k denote discrete time instances for each process). Assuming that the random process $x_i(k)$ is ergodic for all i, we can utilize the time average operator to estimate the correlation function. In this case, the estimate of correlation between two neurons, $r_{ij}(m - k)$, can be obtained by

$$r_{ij}(m-k) = E[x_i(m)x_j(k)] \approx \frac{1}{N-1}\sum_{n=1}^{N} x_i(n-m)x_j(n-k),$$

(3.15)

The cross-correlation vector $\mathbf{p}_{ic}$ can be decomposed and estimated in the same way.

$r_{ij}(m - k)$ is estimated using (3.15) from the neuronal bin count data with $x_i(m)$ and $x_j(k)$ being the bin count of neurons i and j, respectively. From (3.15), it can be seen that $r_{ij}(m - k)$ is equal to $r_{ji}(k - m)$. Because these two correlation estimates are positioned at the opposite side of the diagonal entries of $\mathbf{R}$, the equality leads to a symmetric $\mathbf{R}$. The symmetric matrix $\mathbf{R}$, then, can be inverted effectively by using the Cholesky factorization [28]. This factorization reduces the computational complexity for the inverse of $\mathbf{R}$ from $O(N^3)$ using Gaussian elimination to $O(N^2)$ where N is the number of parameters.

In the application of this technique for BMIs, we must consider the nonstationary characteristics of the input, which can be investigated through observation of the temporal change of the

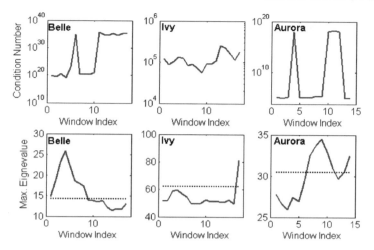

FIGURE 3.3: Illustrations of condition number and nonstationary properties of the input autocorrelation matrix. The dotted lines in the bottom panel indicate the reference maximum eigenvalue that is computed over all the data samples.

input autocorrelation matrix shown in Figure 3.3. The autocorrelation matrix of the multidimensional input data (here 104 neurons) is estimated based on the assumption of ergodicity.[1] In order to monitor the temporal change, the autocorrelation matrix is estimated for a sliding time window (4000-sample length), which slides by 1000 samples (100 second). For each estimated autocorrelation matrix, the condition number and the maximum eigenvalue are computed as approximations of the properties of the matrix. The experimental results of these quantities for three BMI data sets (three different animals) are presented in Figure 3.3. As we can see, the temporal variance of the input autocorrelation matrix is quiet large. Moreover, the inverse of the autocorrelation is poorly conditioned. Notice that in the formulation presented here, $\mathbf{R}$ must be a nonsingular matrix to obtain the optimal solution from (3.10). However, if the condition number of $\mathbf{R}$ is very large, the optimal weight matrix $\mathbf{W}_{\text{Wiener}}$ may be inadequately determined due to round off errors in the arithmetic. This usually happens when the number of samples is too small or the input variables are linearly dependent on each other (as may occur with neural recordings in a small volume of tissue). In such a case, we can reduce the condition number by adding an identity matrix multiplied by some constant to $\mathbf{R}$ before inversion. This procedure is called ridge regression in statistics [31], and the solution obtained

---

[1]A property of some stationary random processes in which the statistical moments obtained by statistical operators equal the moments obtained with time operators.

by this procedure turns out to minimize a cost function that linearly combines the one in (3.13) and a regularization term. This will be addressed in more detail later under regularized solutions.

Provided that the weights of the Wiener filter are adequately trained, one can begin to interpret their meaning as shown in the Hinton diagram in Figure 3.4 for a BMI model trained for a hand-reaching task. Each column of $\mathbf{W}_{\text{Wiener}}$ (i.e., the weight vector of the Wiener filter for each coordinate) is rearranged in a matrix form to show spatiotemporal structure of weight vectors. In this matrix form, the neuron indices are aligned in the x axis and the time lags are in the y axis. Note that the first row of the matrix corresponds to the zero lag (for the instantaneous neuronal bin counts), followed by the successive rows corresponding to the increasing lags (up to nine). In the Hinton diagram, white pixels denote positive signs, whereas black ones do negative signs. Also, the size of pixels indicates the magnitude of a weight. From the Hinton diagram, we can probe the contribution of individual neurons to the output of the Wiener filter. For this purpose, the weights represented in the Hinton diagrams are yielded from the input in which each neuronal bin count time series $x_j(n)$ is normalized to have unit variance. We can see the sign of the correlation between a particular neuronal input and the output. For instance, the weights for neurons indexed by 5, 7, 21, 23, and 71 exhibit relatively large positive values for reaching indicating that those neuronal activities are positively correlated with the output. On the other hand, the weights for neurons 26, 45, 74, and 85 exhibit large negative values indicating the negative correlation between neuronal inputs and the output. There are also some neurons for which the weights have positive and negative values

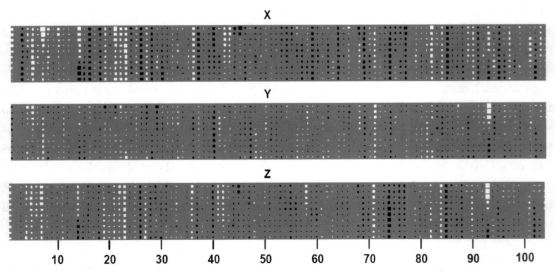

FIGURE 3.4: Hinton diagram of the weights of the Wiener filter for hand reaching.

(e.g., 14 and 93). It is possible from these diagrams to examine the significant time lags for each neuron in terms of the contribution to the filter output. For instance, in the case of neuron 7 or 93, the recent bin counts seem to be more correlated with the current output. However, for neuron 23 or 74, the delayed bin counts seem to be more correlated with the current output. The corresponding reconstruction trajectories for this particular weight matrix are presented in Figure 3.5. For the two reaches presented, the Wiener filter does a fair job at reconstructing the general shape of the trajectory but has a difficult time reaching the peaks and reproducing the details of the movement.

3.1.2 Iterative Algorithms for Least Squares: The Normalized LMS

As is well-known [32], there are iterative or adaptive algorithms that approximate the Wiener solution sample by sample. The most widely known family of algorithms is based on gradient descent, where the weights are corrected at each sample proportionally to the negative direction of the gradient.

$$w(n+1) = w(n) - \eta \nabla J(n) \qquad (3.16)$$

where η is called the stepsize. The most famous is without a doubt the LMS algorithm, which approximates locally the gradient of the cost function yielding

$$w(n+1) = w(n) + \eta e(n)x(n) \qquad (3.17)$$

One of the amazing things about this procedure is that the weight update is local to the weights and requires only two multiplications per weight. It is therefore very easy to program and cuts the complexity of the calculations to $O(N)$. It is therefore very appropriate for DSP hardware implementations.

The price paid by this simplifying alternative is the need to set an extra parameter, the stepsize that controls both the speed of adaptation and the misadjustment, which is defined as the normalized excess MSE (i.e., an added penalty to the minimum MSE obtained with the optimal LS solution). The stepsize is upper bounded by the inverse of the largest eigenvalue of the input autocorrelation matrix for convergence. Practically, the stepsize can be estimated as 10% the inverse of the trace of the input autocorrelation matrix [33]. The steepest descent algorithms with a single stepsize possess an intrinsic compromise between speed of adaptation and misadjustment captured by the relations:

$$M = \frac{\eta}{2} Tr(R) \qquad (3.18)$$

$$\tau = \frac{1}{2\eta \lambda_{min}} \qquad (3.19)$$

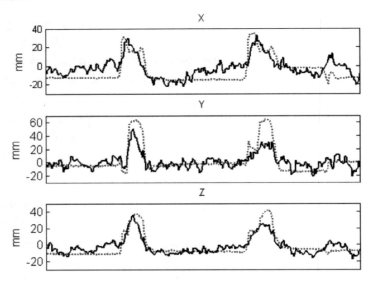

FIGURE 3.5: Wiener filter testing. The actual hand trajectory (dotted red line) and the estimated hand trajectory (solid black line) in the x, y, and z coordinates for a 3D hand-reaching task during a sample part of the test data set.

where τ is the settling time (four times the time constant), $\lambda_{\min}$ is the smallest of the eigenvalues of the input correlation matrix, and M is the misadjustment. Therefore, to speed up the convergence (i.e., less samples to reach the neighborhood of the optimal weight vector), the misadjustment increases proportionally (i.e., the algorithm rattles around the optimal weight value with a larger radius).

Practically, the LMS algorithm has other difficulties, in particular, when the input signal has large dynamical range and the stepsize is chosen for fast convergence. Indeed, in any adaptive algorithm, the weight vector contains all the information extracted from previous data. If the algorithm diverges for some reason, all the past data information is effectively lost. Therefore, avoiding divergence is critical. Since Because the theory for setting the stepsize is based on temporal or statistical averages, an algorithm with a stepsize that converges on average, can transiently diverge. This will happen in times of large errors (difficult to identify) or when the input signal has abrupt large values. To avoid this transient behavior, the normalize stepsize algorithm was developed [2].

The normalized stepsize algorithm is the solution of an optimization problem where the norm of the weight update is minimized subject to the constraint of zero error, that is,

$$J = \left\| w(n+1) - w(n) \right\|^2 + \lambda(d(n) - w(n+1)x(n)) \qquad (3.20)$$

where λ is the Lagrange multiplier and $d(n)$ is the desired response. Solving this optimization problem with Lagrange multipliers, the weight update becomes

$$w(n+1) = w(n) + \frac{1}{\|x(n)\|^2} x(n)e(n) \tag{3.21}$$

Practically, in our BMI linear model, (3.21) becomes

$$\mathbf{w}_{NLMS}^c(n+1) = \mathbf{w}_{NLMS}^c(n) + \frac{\eta}{\gamma + \|\mathbf{x}(n)\|^2} e^c(n)\mathbf{x}(n) \tag{3.22}$$

where η satisfies $0<\eta<2$, and γ is a small positive constant. $e^c(n)$ is an error sample for the c coordinate, and $x(n)$ is an input vector. If we let $\eta(n) \equiv \eta/(\gamma+||\mathbf{x}(n)||^2)$, then the NLMS algorithm can be viewed as the LMS algorithm with a time-varying learning rate such that,

$$\mathbf{w}_{NLMS}^c(n+1) = \mathbf{w}_{NLMS}^c(n) + \eta(n)e^c(n)\mathbf{x}(n) \tag{3.23}$$

Although the weights in NLMS converge in the mean to the Wiener filter solution for stationary data, they may differ for nonstationary data. Therefore, this algorithm may, in fact, have better tracking performance when compared with the Wiener filter applied to segments of the data. Therefore, one should compare its performance with the Wiener filter for BMIs. For the same data set presented in Section 3.1.1, a 10-tap linear filter was trained with LMS (stepsize = 0.0001) using a data set of 10000 pts. To investigate the effect of an online update, the absolute value of the update for each weight was saved at each time t. As shown in Figure 3.6, histogram bins are placed in the range of 0 to 0.01, with centers placed every 0.001. Each coordinate direction for the 3D reaching task has 1040 weights associated with it. Each color in the histogram represents one of the 1040 weights. For example, dark blue represents weight 1, and it was updated 10 000 times, most of which had a value of zero. Any occurrences of an update greater than 0.01 is placed in the 0.01 bin. Only 15% of the total number of weight updates had a value other than zero. For this particular stepsize, this result indicates that one may be able to reduce the number of calculations because many of the weights are not contributing significantly to the mapping. The experiment was also repeated with the normalized LMS algorithm (time-varying stepsize), and the same trend can be observed.

Once the model was trained it was also tested for the trajectories presented in Section 3.1.1, as shown in Figure 3.6. For this BMI application, the performance between the Wiener filter and the filter trained using LMS were essentially the same. The performance was quantified by using the correlation coefficient (CC) between the model output and the true trajectory. Here, the Wiener filter had a CC of 0.76 ± 0.19) while the LMS generated a CC of 0.75 ± 0.20.

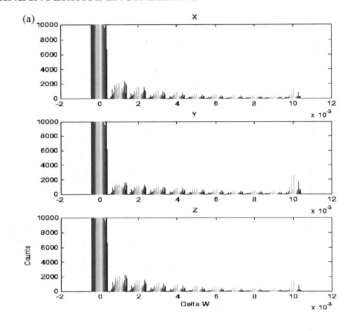

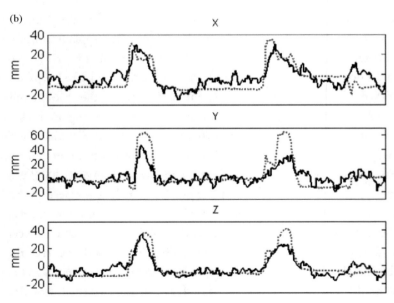

FIGURE 3.6: (a) Histogram of weight updates for the LMS algorithm. LMS training of a linear model. (b) The actual hand trajectory (dotted red line) and the estimated hand trajectory (solid black line) in the x, y, and z coordinate for a 3D hand-reaching task during a sample part of the test data set. (c) Learning curves for LMS training as a function of the stepsize.

(c)

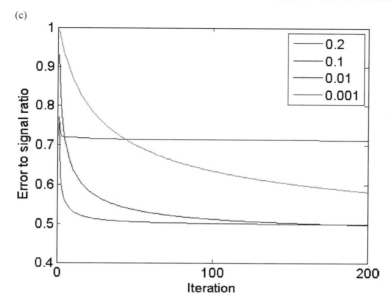

Figure continued

3.2   NONLINEAR MODELS

3.2.1   Time Delay Neural Networks

The primary assumption in the Wiener filter topology is that the input output map is linear, which may not be the case for the neural representation of movement control. To move beyond this assumption for biological systems, dynamic neural networks (a time delay neural network or recurrent neural networks [11, 12, 19, 25, 34, 35]) can be utilized to enhance the simple linear solution at the expense of more sophisticated topologies and training algorithms. In time series analysis, static nonlinear architectures such as the multilayer perceptron (MLP) or the radial basis function network cannot be utilized. Indeed, the hallmark of time series is that there are dependencies hidden in the time structure of the signal, and this holds true for the dynamics of neuronal modulation. Therefore, instantaneous mappers will not capture these dependencies over the lags. Spatiotemporal nonlinear mappings of neuronal firing patterns to hand position can be constructed using an input based time-delay neural networks (TDNNs) [23]. The input based TDNN is an MLP preceded by a delay line that feeds each processing element (PE) of the hidden MLP layer with a vector of present and past input samples. Normally, the nonlinearity is of the sigmoid type (either a logistic

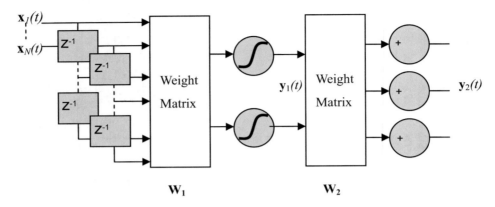

FIGURE 3.7: TDNN topology.

function or a hyperbolic tangent [23]). A full description of the TDNN can be found in References [23, 33]. One of the appeals of the input based TDNN is that it is a dynamic neural network but still utilizes the conventional backpropagation training of the MLP [33].

The TDNN topology (Figure 3.7) is more powerful than the linear FIR filter. In fact, looking at the hidden layer PE just before the nonlinearity, we can recognize an adaptive FIR filter. Therefore, the TDNN is a nonlinear combination of adaptive FIR filters, and has been shown to be a universal approximator in functional spaces [36], provided there is enough memory depth and sufficient number of hidden PEs. Alternatively, each of the hidden PEs outputs can be thought of as an adaptive basis obtained nonlinearly from the high dimensional input data, which defines the projection space for the desired response. Then, the best (orthogonal) projection of the desired hand movements can be obtained on this projection space. Notice that the TDNN can be considered as a "double" adaptive system, where both the axes of the projection space and the optimal projections are adapted. This conceptual picture is important to both help configure the TDNN topology and the training algorithmic parameters.

The output of the first hidden layer of the network can be described with the relation $y_1(t) = f(W_1x(t))$ where $f(.)$ is the hyperbolic tangent nonlinearity $(\tanh(\beta x))$.[2] The input vector x includes L most recent spike counts from N input neurons. In this model the delayed versions of the firing counts, $x(t - L)$, are the bases that construct the output of the hidden layer. The number of delays in the topology should be set that there is significant coupling between the neuronal input and desired signal. The output layer of the network produces the hand trajectory $y_2(t)$ using a linear combination of the hidden states and is given by $y_2(t) = W_2y_1(t)$. The weights (W_1, W_2) of this network can

---

[2]The logistic function is another common nonlinearity used in neural networks.

be trained using static backpropagation[3] with MSE as the learning criterion. In the TDNN static backpropagation algorithm, the output of the first hidden layer is given by (3.24).

$$y_i^l(n) = f(\text{net}_i^l(n)) \tag{3.24}$$

where

$$\text{net}_i^l(n) = \sum_{j=1}^{l} w_{ij}^l y_j^{l-1}(n) \tag{3.25}$$

Here, $y_0^l = x_j$ or the neuronal inputs.

The output layer delta rule is given in (3.26). Note the derivative of output is equal to one for a linear output as used here.

$$\delta_i^L(n) = e_i(n) f'(\text{net}_i^L(n)) \tag{3.26}$$

$$\delta_i^l(n) = f'(\text{net}_i^l(n)) \sum_k \delta_k^{l+1}(n) w_{ki}^{l+1}(n) \text{ (hidden layer delta rule)} \tag{3.27}$$

The overall weight update for the TDNN is given in (3.28)

$$w_{ij}(n+1) = w_{ij}(n) + \eta f'(\text{net}_i(n))(\sum_k e_k(n) f'(\text{net}_k(n)) w_{ki}(n)) y_j(n) \tag{3.28}$$

Although the nonlinear nature of the TDNN may seem an attractive choice for BMIs, putting memory at the input of this topology presents a difficulty in training and model generalization because the high-dimensional input implies a huge number of extra parameters. For example, if a neural ensemble contains 100 neurons with 10 delays of memory, and the TDNN topology contains 5 hidden PEs, 5000 free parameters are introduced in the input layer alone. Large data sets and slow learning rates are required to avoid overfitting [24]. Untrained weights can also add variance to the testing performance thus decreasing accuracy.

As in any neural network topology, the issues of optimal network size, learning rates, stopping criterion need to be addressed for good performance. There are well-established procedures to control each one of these issues [24]. The size of the input layer delay line can be established from the knowledge of the signal structure. In BMIs, it is well established that one second of neural data preceding the motor event should be used [8]; however, the optimal embedding and the optimal delay should still be established using the tools from dynamic modeling. The rule of thumb to build the architecture is to always start simple (i.e., the smallest number of hidden PEs, which as we saw defines the size of the projection space). As far as training is concerned, one has to realize that both the bases and

---

[3]Backpropagation is a simple application of the chain rule that propagates the gradients through the topology.

the projection onto the projection space are being adapted at the same time. Another rule of thumb is to adapt the bases at a slower rate than the projection, which means that the learning rate of the first layer weights should be smaller than of the second layer weights (even after recognizing that the errors are attenuated by passing through the hidden PE nonlinearity) and to train the network with early stopping criterion (i.e., set aside 10% of the data for a cross-validation (CV) set, and stop training when the error in the CV increases. The error in the CV set should always be comparable to the error in the training set, otherwise, overtraining occurred and generalization will be compromised.

Here, we present the training of a TDNN for the BMI data described in Section 3.1.1. The networks implemented consisted of one hidden layer with a hyperbolic tangent nonlinearity $(\tanh(\beta x))$. Because the dynamic range of the input data was large (0–21), the slope of the nonlinearity had to be carefully chosen. A beta value of 0.5 provided a sufficiently steep slope, which with slow learning (over many epochs) nonlinearities would saturate to produce flat regions in the trajectory. The slope was also shallow enough to prevent instantaneous jumps in firing rate from saturating the nonlinearity and decreasing learning. The output layer of the network consisted of a linear PE because the desired signal was hand position in space. The network described was trained using static backpropagation. Weights were updated in online mode. Because the data have a large variability, the weights needed to be updated at each time instance; batch updates tend to average out the updates making learning difficult for this problem.

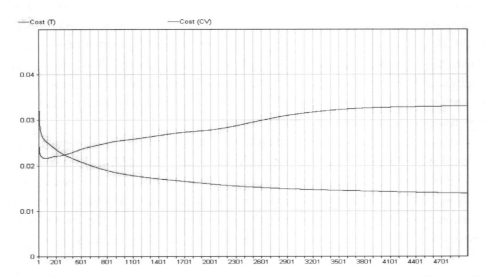

FIGURE 3.8: Learning curve for a BMI TDNN. Here, the y axis relates the MSE to the number of training epochs. Included in the figure is the CV curve. Theoretical analysis indicates that training should be stopped at the minima of the CV curve.

The size of the topology was chosen by experimentation: 3, 5, 15, and 30 PEs were chosen for the hidden layer. Because we seek a model that accurately maps all novel neuronal firing patterns to hand position, we must search for a network topology that minimizes the output bias and variance. The operator must choose learning rates that slowly adjust the weights so that maximum information is captured and the network does not simply track the trajectory. Overfitting the network should also be avoided. The training should be stopped at the point of maximum generalization, which is based on the minimum error produced by a validation set of data [33]. A plot of the CV over epochs is presented in Figure 3.8. For neuronal data, the early stopping method indicates that training should be stopped at approximately 50 epochs of data. This method did not produce good models. Research has verified CVs poor performance in some problems [24]. Other methods will be presented later to quantify maximum generalization for neuronal modeling.

The testing outputs for the TDNN BMI are presented in Figure 3.9. The TDNN performs better than the linear filter in peak accuracy (CC = 0.77 ± 0.17) but suffers in smoothness. The large number of free parameters in the TDNN is not only difficult to train but produce outputs with large variances. The biggest difficulty with BMI models of this type is the enormous set of free parameters produced by the large neural input vector (104 dimensions) and the multiplication by the number of hidden PEs. Training the first layer of the TDNN with backpropagation is a difficult task because of the attenuation of the error by the hidden layer PEs. Decreased input data dimensionality will limit the number of free parameters in the adaptive system architecture. Simulations express the need for preprocessing which prune neurons with the most information in the sparse data set. This problem calls for an adaptive system with the nonlinear power of a TDNN but with a

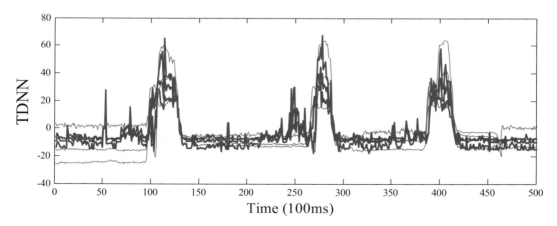

FIGURE 3.9: Testing performance for a TDNN for a reaching task. Here, the red curves are the desired *x*, *y*, and *z* coordinates of the hand trajectory, whereas the blue curves are the model outputs.

fewer number of free parameters at the input layer. Finally, a test for generalization is crucial to the successful production of useful models.

3.2.2 Recursive MLPs

Here, we continue evaluation of nonlinear dynamical models for BMIs by introducing the recursive MLP (RMLP). First, we would like to discuss in detail the list below indicating why the RMLP is an appropriate choice for BMI applications.

- RMLP topology has the desired biological plausibility needed for BMI design
- The use of nonlinearity and dynamics gives the RMLP a powerful approximating capability

Although the RMLP may first appear to be an "off-the-shelf" black box model, its dynamic hidden layer architecture can be compared with a mechanistic model of motor control proposed by Todorov [18, 37], revealing that it has the desired biological plausibility needed for BMI design. The formulation for the RMLP is similar to a general (possibly nonlinear) state space model implementation that corresponds to the representation interpretation of Todorov's model for neural control. The RMLP architecture in Figure 3.10 consists of an input layer with N neuronal input channels, a fully connected hidden layer of nonlinear PEs, (in this case tanh), and an output layer of linear PEs, one for each output. The RMLP has been successfully utilized for difficult control applications [38], although here the more traditional backpropagation-through-time (BPTT) training [39] was utilized.

Each hidden layer PE is connected to every other hidden PE using a unit time delay. In the input layer equation of (3.29), the state produced at the output of the first hidden layer is a nonlinear function of a weighted combination (including a bias) of the current input and the previous

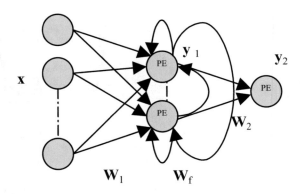

FIGURE 3.10: Fully connected, state recurrent neural network.

state. The feedback of the state allows for continuous representations on multiple timescales and effectively implements a short-term memory mechanism of the state values. Here, $f(.)$ is a sigmoid nonlinearity (in this case, tanh), and the weight matrices $\mathbf{W}_1$, $\mathbf{W}_2$, and $\mathbf{W}_f$, as well as the bias vectors $\mathbf{b}_1$ and $\mathbf{b}_2$ are again trained using synchronized neural activity and hand position data. Again, as in the TDNN, each of the hidden PEs outputs can be thought of as a nonlinear adaptive basis partially created from the input space and utilized to project the high-dimensional input data. These projections are then linearly combined to form the outputs of the RMLP that will predict the desired hand movements as shown in (3.30). It should be now apparent the elegance of feedback in decreasing the number of free parameters when compared with the input-based TDNN. Indeed, the state feedback in the RMLP allows for an input layer weights defined by the number of inputs while requiring only $K^2\mathbf{W}_f$ weights, where K is the number of hidden PEs for a total of $N + 2K^2$ total weights per output. One of the disadvantages of the RMLP when compared with the TDNN is that the RMLP cannot be trained with backpropagation and requires either BPTT or real-time recurrent learning [33]. Either of these algorithms is much more time-consuming than standard backpropagation, and on top of this, the dynamics of learning are harder to control with the stepsize, requiring extra care and more experience for good results.

$$\mathbf{y}_1(t) = f\left(\mathbf{W}_1\mathbf{x}(t) + \mathbf{W}_f\mathbf{y}_1(t-1) + \mathbf{b}_1\right), \tag{3.29}$$

$$\mathbf{y}_2(t) = \mathbf{W}_2\mathbf{y}_1(t) + \mathbf{b}_2, \tag{3.30}$$

The performance of RMLP input–output models in BMI experiments is dependent upon the choice of the network topology, learning rules, and initial conditions. Therefore, a heuristic exploration of these settings is necessary to evaluate the performance.

The first consideration in any neural network implementation is the choice of the number of PEs. Because the RMLP topology studied here always consisted of only a single hidden layer of tanh PEs, the design question becomes one of how many PEs are required to solve the neural to motor mapping. The first approach to this problem was a brute-force scan of the performance across a range of PEs, as shown in Table 3.1. It can be seen for a reaching task that the topology with five PEs produced the highest testing correlation coefficients. Although the brute-force approach can

TABLE 3.1: RMLP performance as a function of the number of hidden PEs

	1 PE	5 PE	10 PE	20 PE
Average CC testing	0.7099	0.7510	0.7316	0.6310

immediately direct the choice of topology, it does not explain why the number is appropriate for a given problem. We later came to find out that the number of hidden PEs should be chosen to span the space of the desired trajectory [34]. In the context of BMIs, this knowledge can help to avoid the computational and time costs of the brute-force approach.

The RMLP presented here was trained with BPTT [33, 39] using the NeuroSolutions software package [40]. Training was stopped using the method of CV (batch size of 1000 points) (Figure 3.11) to maximize the generalization of the network [41]. The BPTT training procedure involves unfolding the recurrent network into an equivalent feedforward topology over a fixed interval, or trajectory. For our BMI applications, we are interested in learning the dynamics of the trajectories of hand movement; therefore, for each task, the trajectory was chosen to match a complete movement. For the reaching task, this length was, on average, 30 samples. In Table 3.2, this selection of trajectory was compared again with a brute-force scanning of the testing performance as a function of the trajectory length (samples per exemplar). Included in the BPTT algorithm is the option to update the weights in online (every trajectory), semibatch, or batch mode. Choosing to update the network weights in semibatch mode can protect against noisy stochastic gradients that can cause the network to become unstable by averaging the gradients over several trajectories. Table 3.2 also provides the average testing correlation coefficient as a function of the frequency of updates. We can see that a range of good choices [15–30 samples/exemplar (s/e) and 5–15 exemplar/update (e/u)] exist for the trajectory length and update rule.

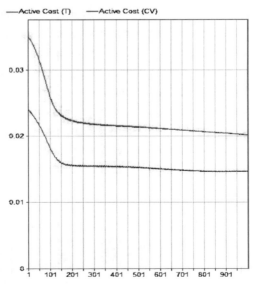

FIGURE 3.11: RMLP learning curve. MSE (upper curve) and CV (lower curve).

TABLE 3.2: Average testing correlation coefficients as a function of trajectory length

	5 S/E	15 S/E	30 S/E	60 S/E
1 e/u	0.6514	0.7387	0.7302	0.7135
5 e/u	0.6458	0.7006	0.7263	0.7091
15 e/u	0.6544	0.7389	0.6680	0.6855
30 e/u	0.6807	0.7482	0.7116	0.6781
60 e/u	0.6457	0.6383	0.6772	0.6624

To test the effect of the initial condition on model performance, 100 Monte Carlo simulations with different initial conditions were conducted with 20 010 consecutive bins (2001 sec) of neuronal data to improve the chances of obtaining the global optimum. In Figure 3.12, the training MSE is presented for all simulations, and it can be seen that all initial conditions reach the approximately the same solution. The greatest effect of the initial condition can be seen in the time it takes for each model to converge. In Figure 3.13, the average and standard deviations of the curves in Figure 3.12

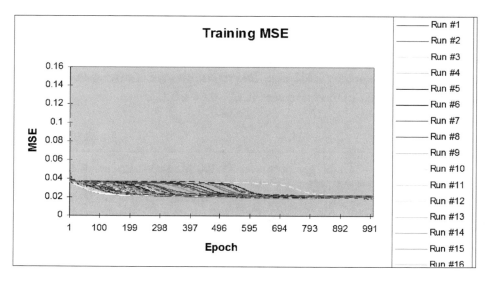

FIGURE 3.12: Training MSE curves for 100 Monte Carlo simulations.

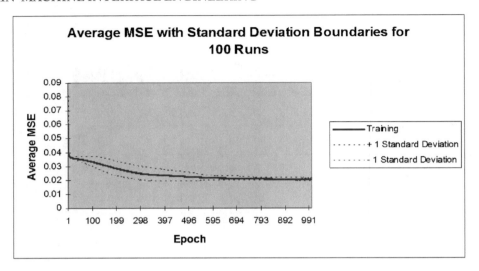

FIGURE 3.13: Standard deviation in training MSE for 100 Monte Carlo simulations.

are presented. This figure is characterized by a large standard deviation in the initial epochs resulting from the initial condition and a small standard deviation in the final epochs signifying that the
models are converging to the same solution. These relationships are quantified in Table 3.3 where
the average final MSE had a value of 0.0203 ± 0.0009. Again, a small training standard deviation
indicates the networks repeatedly achieved the same level of performance. Of all the Monte Carlo
simulations, the network with the smallest error achieved an MSE of 0.0186 (Table 3.4).

   We can see in the testing outputs for the RMLP shown in Figure 3.14 that the performance
of a parsimonious nonlinear dynamical network can improve the testing performance for trajectory
reconstruction in a hand-reaching BMI task. The trajectories are clearly smoother and distinctly
reach each of the three peaks in the movement (CC = 0.84 ± 0.15).

TABLE 3.3: Training performance for 100 Monte Carlo simulations		
ALL RUNS	**TRAINING MINIMUM**	**TRAINING STANDARD DEVIATION**
Average of minimum MSEs	0.020328853	0.000923483
Average of final MSEs	0.020340851	0.000920456

TABLE 3.4: Best performing network	
BEST NETWORK	**TRAINING**
Run	85
Epoch	999
Minimum MSE	0.018601749
Final MSE	0.018616276

3.2.2.1 Echo-State Networks. The improvement in performance of the RMLP is gained at the expense of a significant leap in the computational costs of training. The problem is accentuated by the fact that these models may have to be retrained occasionally to adjust to the changing statistics and environment. This can severely restrict practical, real-world application of BMIs. We show here a different architecture that has performance similar to that of an RMLP, but the training has linear complexity as opposed to the (BPTT) RMLP training algorithm [39].

Echo state networks (ESNs) were first proposed by Jaeger [42] and are appealing because of their computational abilities and simplified learning mechanisms. ESNs exhibit some similarities to the "liquid state machines" proposed by Maas et al. [43], which possess universal approximation

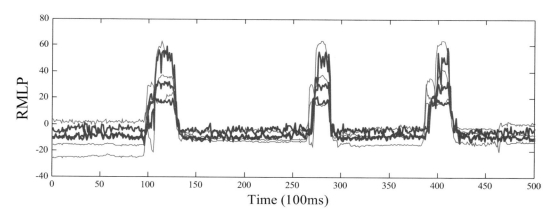

FIGURE 3.14: Testing performance for a RMLP for a reaching task. Here, the red curves are the desired *x*, *y*, and *z* coordinates of the hand trajectory, whereas the blue curves are the model outputs.

capabilities in myopic functional spaces. The fundamental idea of an ESN is to use a "large-reservoir" recurrent neural network that can produce diversified representations of an input signal, which can then be instantaneously combined in an optimal manner to approximate a desired response. Therefore, ESNs possess a representational recurrent infrastructure that brings the information from the past of the input into the present sample without adapting parameters. The short-term memory of the ESNs is designed a priori, and the neuronal data are processed through a reservoir of recurrent networks that are linked by a random, sparse matrix. Therefore, only the adaptive linear or nonlinear regressor (static mapper) is needed to implement functional mappings. Therefore, the training can be done online in $O(N)$ time, unlike the Wiener or the RMLP that have $O(N^2)$ training algorithms. Figure 3.15 shows the block diagram of an ESN. A set of input nodes denoted by the vector $\mathbf{u}_n \in \mathfrak{R}^{M \times 1}$ is connected to a "reservoir" of N discrete time-recurrent networks by a connection matrix $\mathbf{W}_{in} \in \mathfrak{R}^{M \times N}$. At any time instant n, the readout (state output) from the recurrent neural network (RNN) reservoir is a column vector denoted by $\mathbf{x}_n \in \mathfrak{R}^{N \times 1}$. Additionally, an ESN can have feedback connections from the output to the RNN reservoir. In Figure 3.15, we show two outputs (representing the X–Y Cartesian coordinates of hand trajectory) and the associated feedback connection matrix, $\mathbf{W}_b \in \mathfrak{R}^{N \times 2} = \left[\mathbf{w}_{b1}, \mathbf{w}_{b2}\right]$. The desired outputs form a 2D column vector $\mathbf{d}_n = [d_{xn}; d_{yn}]$. The reservoir states are transformed by a static linear mapper that can additionally receive contributions from the input $\mathbf{u}_n$. Each PE in the reservoir can be implemented as a leaky integrator (first-order gamma memory [44]) and the state output or the readout is given by the difference equation

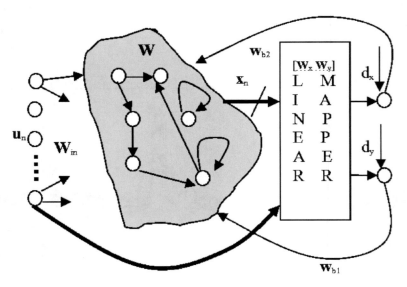

FIGURE 3.15: Block diagram of the echo state networks.

$\mathbf{x}_{n+1} = (1 - \mu\, Ca)\mathbf{x}_n + \mu C(f(\mathbf{W}_{in}^T \mathbf{u}_{n+1} + \mathbf{W}\mathbf{x}_n + \mathbf{W}_b \mathbf{d}_n))$, where $0 < \mu < 1$ is the step-size used for converting a continuous-time leaky integrator into a discrete time difference equation, C is the time constant, and a is the decay rate [42]. The point-wise nonlinear function $f(.)$ is chosen to be the standard tanh sigmoid (i.e., $f(.)$ = tanh$(.)$). Note that, if μ, C, and a are all equal to unity, the RNNs default to the conventional nonlinear PE [33] without memory. From a signal processing point of view, the reservoir creates a set of bases functions to represent the input, whereas the static mapper finds the optimal projection in this space. There are obvious similarities of this architecture to kernel machines, except that the kernels here are time functions (Hilbert spaces).

We will now give the conditions under which an ESN can be "useful," which Jaeger aptly calls as the "echo state property." Loosely stated, the echo state property says that the current state is uniquely defined by the past values of the inputs and also the desired outputs if there is feedback. A weaker condition for the existence of echo states is to have the spectral radius of the matrix $\hat{\mathbf{W}} = \mu C \mathbf{W} + (1-\mu Ca)\mathbf{I}$ less than unity [42]. Another aspect critical for the success of ESN is to construct a sparse matrix $\mathbf{W}$. This will ensure that the individual state outputs have different representations of the inputs and desired outputs, or in other words, the span of the representation space is sufficiently rich to construct the mapping to the desired response (hand trajectory).

3.2.2.2 Design of the ESN. One of the important parameters in the design of the ESN is the number of RNN units in the reservoir. In our experience with BMI simulations, Monte Carlo simulations varying the number of parameters and monitoring performance provided the best technique for selecting the dimensionality of the reservoir. Here, we chose N = 800, where increasing N further did not result in any significant improvement in performance. The input weight matrix $\mathbf{W}_{in}$ was fully connected with all the weights fixed to unity. The recurrent connection matrix $\mathbf{W}$ was sparse, with only 1% of the weights (randomly chosen) being nonzero. Moreover, we fix all the nonzero weights to a value 0.5. Further, each RNN is a gamma delay operator with parameters $[a, C, \mu] = [1, 0.7, 1]$. The next aspect is to set the spectral radius, which is crucial for this problem because it controls the dynamics and memory of the echo states. Higher values are required for slow output dynamics and vice versa [42, 45]. For the experiments in this section, we utilized a single ESN whose spectral radius was tuned to 0.79. Marginal changes (<1%) in performance (both X and Y) were observed when this parameter was altered by ±10%. We also turned off the connections from the output to the RNN reservoir and the direct connections between the inputs and the linear mapper. The network state is set to zero initially. The training inputs were forced through the network, and the states were updated. The first 400 echo state outputs were discarded as transients. The remaining 5000 state outputs were used to train the linear mapper.

Testing of the network on a 2D cursor control task demonstrated that the ESN can perform at the same level as the standard RMLP trained with BPTT. The CC values for the ESN x and y coordinates, respectively, were (0.64 and 0.78), whereas the RMLP produced CC values of (0.66 and 0.79). However, the case of the ESN was trained with far less complexity.

Lastly, we would like to mention the appeal of the ESN as a model for biologically plausible computation. If one thinks of the echo states as neuronal states, we can see how a distributed, recurrent topology is capable of storing information about past inputs into a diffuse set of states (neurons). For the most part, the interconnectivity and the value of the weights (synapses) are immaterial for representation. They become, however, critical for readout (approximation). In this respect, there are very close ties between ESN and liquid state machines, as already mentioned by Maas et al. [43]. These ideas may become useful in developing new distributed paradigms for plasticity and characterization of neuronal response in motor cortex.

3.2.3 Competitive Mixture of Local Linear Models

So far, we have investigated nonlinear decoding models that globally approximate an unknown nonlinear mapping between neural activities and behavior. However, a complex nonlinear modeling task can be elucidated by dividing it into simpler tasks and combining them properly [46]; this is an application of the well-known divide-and-conquer approach extensively used in science and engineering. We will briefly review here the statistical framework called *mixture of experts* [47] that implements this approach, and will also extend it to a more general model called *gated competitive experts* that includes other neural architectures that have been successfully used in time series segmentation [48] and optimal control [49]. We will here summarize the methodology that was originally presented in Reference [48].

Let us consider modeling a vector i.i.d.[4] process that may be multimodal, assuming that the mixing process has memory using a mixture model. To simplify the derivation, a 1D time series is assumed, but the results can easily be generalized to the multidimensional case. Recall that the multivariate PDF of a random process, with an effective memory depth of M, can be decomposed as

$$p(\chi(n)) = \prod_{n=1}^{N} p(x(n) \mid \chi(n-1)) \qquad (3.31)$$

where $\chi(n - 1) = [x(n - 1), \ldots x(n - M+1)]^{\mathrm{T}}$. Let us entertain the possibility that the random process is produced by K switching regimes. Therefore, we propose the joint conditional density $p(k, x(n) \mid \chi(n-1))$, where the discrete variable $k = 1 \ldots K$ indicates the regime. We cannot observe

---

[4]Independent identically distributed.

this PDF directly because we do not know which regime is active at which times, but we can observe its marginal distribution

$$p(x(n)|\chi(n-1)) = \sum_{k=1}^{K} p(k, x(n)|\chi(n-1)) =$$

$$\sum_{k=1}^{K} P(k|\chi(n-1)) \cdot p(x(n)|\chi(n-1), k) = \sum_{k=1}^{K} g_k(\chi(n-1)) \cdot p(x(n)|\chi(n-1), k) \qquad (3.32)$$

where $g_k(\chi(n-1)) \equiv P(k|\chi(n-1))$ can be regarded as the a priori probability of the kth regime. If we can find predictors for each of the subprocesses, up to a random term ε_k,

$$x(n) = f(\chi(n-1); \theta_k) + \varepsilon_k(n) \qquad k = 1 \dots K \qquad (3.33)$$

then we can evaluate the conditional PDF in terms of the innovations

$$p(x(n)|\chi(n-1), k) = p_{\varepsilon_k}(\varepsilon_k(n)) \qquad (3.34)$$

The residuals are usually modeled as a Gaussian distribution, which for a 1D time series becomes

$$p_{\varepsilon_k}(\varepsilon_k) = \frac{1}{\sqrt{2\pi\sigma_k^2}} \exp\left[\frac{(\tilde{x}_k(n) - x(n))^2}{2\sigma_k^2}\right] \qquad (3.35)$$

where we have $\tilde{x}_k(n) \equiv f(\chi(n-1); \theta_k)$ defined to be kth predictor's estimate of the next value of the time series and σ_k, a "nuisance parameter," is the variance of the kth predictor. Taking the expected value of both sides of Eq 3.33 gives the best MMSE prediction of the next value of the time series

$$\tilde{x}(n) \equiv E[x(n)|\chi(n-1)] = \sum_{k=1}^{K} g_k(\chi(n-1)) \cdot E(x(n)|\chi(n-1), k] = \sum_{k=1}^{K} g_k(\chi(n-1))\tilde{x}_k(n) \qquad (3.36)$$

which is a weighted sum of outputs of the individual predictors. We can regard this as the total system output. In the mixture-of-experts algorithm of Jordan and Jacobs [50], this mapping is provided by an adaptable function called the "gate." The particular variation of the algorithm we will examine is Weigend's [51] nonlinear gated experts, which implements the gate using a single MLP. This architecture is shown in Figure 3.16, and note that the individual predictors, called "experts" in this context, and the gates all see the same input. In other implementations, the gate can be trained from the outputs (called therefore output gating).

We now turn to the question of how to train the gate and experts simultaneously, which is the great appeal of this architecutre. In the following development, for ease of reading, we leave out

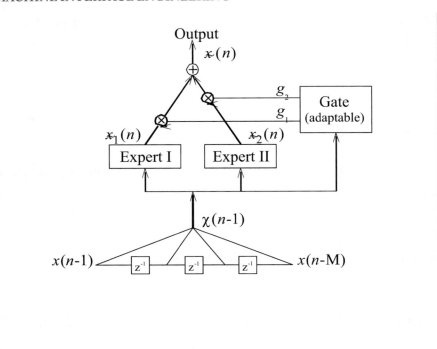

FIGURE 3.16: Mixture of experts during activation for a network with two experts.

explicit time dependence whenever possible. Also, there is an implicit iteration index in all the following equations. Given a time series of length N, we choose the free parameters of the predictors and gate that maximize the process log-likelihood. If the innovations are i.i.d., we can rewrite the process likelihood as

$$p(\chi(n)) = \prod_{n=1}^{N} \sum_{k=1}^{K} g_k(\chi(n-1)) \cdot p_{\varepsilon_k}(\tilde{x}_k(n) - x(n)) \qquad (3.37)$$

This is difficult to maximize directly. Therefore, we propose a latent binary indicator, I_k, indicating which expert is valid, allowing the likelihood to be written.

$$L = \prod_{n=1}^{N} \prod_{k=1}^{K} [g_k(\chi(n-1)) \cdot p(x(n) \mid \chi(n-1), k)]^{I_k(n)} \qquad (3.38)$$

The indicator variable is "hidden," in the sense that we do not know a priori which expert is valid at any time step. In the E step of the EM algorithm, for a given set of free parameters of the experts and gate, the entire data set is evaluated, holding the free parameters constant, to determine $p(x(n) \mid \chi(n-1), k)$ and $g_k(\chi(n-1))$ for all k and n. We then replace the indicator variables, I_k, at every time step, by their expected value:

$$h_k(n) \equiv E[I_k(n)] = P(k \,|\, x(n), \chi\,(n-1)) =$$

$$\frac{P(k \,|\, \chi\,(n-1)) \cdot p(x(n) \,|\, \chi\,(n-1), k)}{p(x(n) \,|\, \chi\,(n-1))} = \frac{g_k(\chi\,(n-1)) \cdot p(x(n) \,|\, \chi\,(n-1), k)}{\displaystyle\sum_{k=1}^{K} g_k(\chi\,(n-1)) p(x(n) \,|\, \chi\,(n-1), k)} \qquad (3.39)$$

Thus, h_k is the posterior probability of expert k, given both the current value of the time series and the recent past. For the M step, L is maximized or equivalently, the negative log-likelihood,

$$J = \sum_{n=1}^{N} \sum_{k=1}^{K} \left\{ -h_k(n) \cdot \log[\,g_k(\chi\,(n-1))] + \frac{1}{2} h_k(n) \Big[\frac{(\tilde{x}_k(n) - x(n))^2}{\sigma_k^2} + \log[\sigma_k^2] \Big] \right\} \qquad (3.40)$$

is globally minimized over the free parameters. The process is then repeated. If, in the M step, J is only decreased and not minimized, then the process is called the generalized EM algorithm. This is necessary when either the experts or gate is nonlinear, and a search for the global minimum is impractical.

The first term in the summation of (3.40) can be regarded as the cross-entropy between the posterior probabilities and the gate. It has a minimum when only one expert is valid and thus encourages the experts to divide up the input space. To ensure that the outputs of the gate sum to unity, the output layer of the MLP has a "softmax" transfer function,

$$g_k(\chi) = \frac{\exp[s_k(\chi)]}{\displaystyle\sum_{k=1}^{K} \exp[s_j(\chi)]} \qquad (3.41)$$

where s_k is the kth input to the softmax. For a gate implemented as an MLP, the cross entropy term in (3.40) cannot be minimized in a single step, and the generalized EM algorithm must be employed. If the gate is trained through gradient descent (backpropagation), the error backpropagated to the input side of the softmax, at each time step is

$$\frac{\partial J}{\partial s_k} = g_k(\chi) - h_k \qquad (3.42)$$

This is the same backpropagated error that would result for a MSE with the posterior probabilities acting as the desired signal. Thus, the posterior probabilities act as targets for the gate. For each EM iteration, several training iterations may be required for the gate because it is implemented using a multilayer perceptron.

There is an analytical solution for the experts, at each iteration, when they are linear predictors, $\tilde{x}_k(n) = w_k^T \chi\,(n-1)$ and $w_k = \mathbf{R}_k^{-1}$ and $\mathbf{p}_k$, where $\mathbf{R}_k$ and $\mathbf{p}_k$ are weighted autocorrelation and cross-correlation matrices, respectively,

$$R_k = \frac{1}{\sum\limits_{n=1}^{N} h_k(n)} \sum_{n=1}^{N} h_k(n)\chi(n-1)\chi^T(n-1) \tag{3.43}$$

$$P_k = \frac{1}{\sum\limits_{n=1}^{N} h_k(n)} \sum_{n=1}^{N} h_k(n)\chi(n-1)x(n) \tag{3.44}$$

Note that the solution represents a weighted Wiener filter, where the data are weighted by the posterior probabilities. If the experts are nonlinear, then the error used to iteratively train each expert is simply interpreted as being weighted by the posterior probabilities, $h_k(n)$. However, Zeevi et al. [52] have shown that the nonlinearity of the gate allows the mixture of experts to be a universal approximator, even if the expert predictors are linear.

No matter whether the experts are linear or nonlinear, there is an exact solution for the Gaussian covariance parameter σ_k at the end of each epoch

$$\sigma_k^2 = \frac{1}{\sum\limits_{n=1}^{N} h_k(n)} \sum_{n=1}^{N} h_k(n)[\tilde{x}(n) - x(n)]^2 \tag{3.45}$$

3.2.3.1 Gated Competitive Experts.
A gated competitive expert model is characterized by several adaptable "experts" that compete to explain the same data. That is, they all see the same input, and all attempt to produce the same output. Their performance is both monitored and mediated by a gate, whose goal is to determine the relative validity of the experts for the current data and then to appropriately moderate their learning. The history of the gate's decisions can be used to segment the time series.

In the activation phase, the total system output is a simple weighted sum of the experts' outputs

$$y(n) = \sum_{k=1}^{K} g_k(n) \cdot y_k(\chi(n)) \tag{3.46}$$

where y_k is the output of the kth expert with χ as input and g_k is the kth output of the gate. The input, $\chi(n)$ is a vector of delayed versions of the time series, in essence the output of a tapped delay line, repeated here for convenience

$$\chi(n) = \left[x(n)\ x(n-1)\ \dots\ x(n-M+1)\right]^T \tag{3.47}$$

For the mixture to be meaningful, the gate outputs are constrained to sum to one.

$$\sum_{k=1}^{K} g_k(n) = 1 \tag{3.48}$$

Such a linear mixture can represent either a competitive or cooperative system, depending on how the experts are penalized for errors, as determined by the cost function. In fact, it was in the context of introducing their mixture of experts model that Jacobs et al.[53] first presented a cost function that encourages competition among gated expert networks, which we generalize to

$$J(n) = \sum_{k=1}^{N} g_k(n) \cdot f(d(n) - y_k(\chi(n))) \tag{3.49}$$

where d is the desired signal and $f(d(n) - y_k(\chi(n)))$ is a function of the error between the desired signal and the kth expert. Because the desired signal is the same for all experts, they all try to regress the same data and are always in competition. This alone, however, is not enough to foster specialization. The gate uses information from the performance of the experts to produce the mixing coefficients. There are many variations of algorithms that fall within this framework. Let us discuss the important components one at a time, starting with the design of the desired signal.

The formalism represented by (3.49) is a supervised algorithm, in that it requires a desired signal. However, we are interested in a completely unsupervised algorithm. A supervised algorithm becomes unsupervised when the desired signal is a fixed transformation of the input: $d \Rightarrow d(\chi)$. Although many transformations are possible, the two most common transformations involve the delay operator and the identity matrix, resulting in prediction as explained above and auto-association, which yields a generative model for PCA [54].

Gates can be classified into two broad categories, which we designate as input or output based. With input-based gating, the gate is an adaptable function of the input, $g_k \Rightarrow g_k(\chi)$, that learns to forecast which expert will perform the best, as we have seen in the mixture of experts. For output-based gating, the gate is a directly calculated function of the performance, and hence, the outputs, of the experts. The gate in the annealed competition of experts of Pawelzik et al. [55] implements memory in the form of a local boxcar average squared error of the experts. The self-annealing competitive prediction of Fancourt and Principe also uses the local squared error, but using a recursive estimator. The mixture of experts can also keep track of past expert performance, the simplest example of which is the mixture model where the gate is expanded with memory to create an estimate of the average of the posterior probabilities over the data set [51]. Perhaps, the simplest scheme is hard competition, for which the gate chooses the expert with the smallest magnitude error in a winner-take-all fashion, as will be explained below. This method simplifies the architecture and is very appropriate for system identification in a control framework because it simplifies the design of controllers.

The total cost function can be constructed in several ways from the instantaneous cost, depending on the error function $f(\cdot)$. If $f(\cdot)$ is a quadratic function of the error, the total cost over a data set of N samples is the sum of the instantaneous costs.

$$J = \sum_{n=1}^{N}\sum_{k=1}^{N} g_k(n)\cdot[d(n) - y_k(\chi(n))]^2 \qquad (3.50)$$

$$J = \prod_{n=1}^{N}\sum_{k=1}^{N} g_k(n)\cdot p(d(n) - y_k(\chi(n))) \qquad (3.51)$$

If $f(\cdot)$ is a PDF in the error, the appropriate total cost function is a product of the instantaneous costs and returns us toward the mixture of experts when the gate is input based. We explain now in more detail the hard competitive structure (shown in Figure 3.17). The experts, in this case predictors, are located on a spatial grid, analogous to a Kohonen's self-organizing map [56], and the gate moderates the learning of the experts based on their distance from the best performing expert. The best predictor is the one with the smallest local mean squared error

$$\text{winner}(n) = \arg\min_k[\varepsilon_k(n)] \qquad (3.52)$$

where the local MSE is computed using the recursive estimate for each expert, repeated here for convenience

$$\varepsilon_k(n) = \lambda e_k^2(n) + (1-\lambda)\varepsilon_k(n-1) \qquad 0 < \lambda < 1 \qquad (3.53)$$

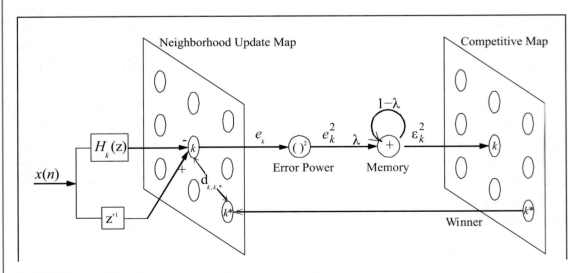

FIGURE 3.17: Neighborhood map of competing predictors.

where e_k is the instantaneous squared error of the kth predictor. The memory term, λ, is identical for all experts, but can be manually adjusted during training. The gating function, which moderates the learning rate of the predictors, is determined by the distance from the winning predictor, k^*, to the other predictors.

$$g_k(n) = \exp\left[\frac{d^2_{k,k^*}(n)}{2\sigma^2(n)}\right] \qquad (3.54)$$

where k is the predictor to be updated, k^* is the winning predictor, d_{k,k^*} is the neighborhood distance between them, and σ is an annealing parameter that controls the neighborhood width. This results in the model shown in Figure 3.17. The signal flow graph is shown for only one predictor.

In exact analogy with training a Kohonen map, both the neighborhood width and the global learning rate are annealed during training according to an exponentially decreasing schedule

$$\sigma(t) = \sigma_0 e^{\frac{-i}{\tau}} \qquad \mu(t) = \mu_0 e^{\frac{-i}{\tau}} \qquad (3.55)$$

where τ is the annealing rate and i is the iteration number. The overall learning rate for the kth predictor at the ith iteration is

$$\eta_k(i) = \mu(i) \cdot g_k(i) \qquad (3.56)$$

3.2.3.2 Gated Competitive Experts in BMIs.

Let us see how the hard competition of gated experts can be applied to BMIs. Hypothesizing that the neural activity will demonstrate varying characteristics for different localities in the space of the hand trajectories, we expect the multiple model approach, in which each linear model specializes in a local region, to provide a better overall input–output mapping. However, the BMI problem is slightly different from the signal segmentation because the goal is to segment the joint input/desired signal space.

The training of the multiple linear models is accomplished by competitively (hard or soft competition) updating their weights in accordance with previous approaches using the NLMS algorithm. The winning model is determined by comparing the (leaky) integrated squared errors of all competing models and selecting the model that exhibits the least integrated error for the corresponding input [48]. The leaky integrated squared error for the ith model is given by

$$\varepsilon_i(n) = (1 - \mu)\varepsilon_i(n-1) + \mu e_i^2(n), i = 1, \cdots, M, \qquad (3.57)$$

where M is the number of models and μ is the time constant of the leaky integrator. Then, the jth model wins competition if $\varepsilon_j(n) < \varepsilon_i(n)$ for all $i \neq j$. If hard competition is employed, only the weight vector of the winning model is updated. Specifically, if the jth model wins competition, the update rule for the weight vector $\mathbf{w}_j(n)$ of that model is given by

$$\mathbf{w}_j(n+1) = \mathbf{w}_j(n) + \eta \, \frac{e_j(n)\mathbf{x}(n)}{\gamma + \|\mathbf{x}(n)\|^2},$$ (3.58)

where $e_j(n)$ is the instantaneous error and $\mathbf{x}(n)$ is the current input vector. η represents a learning rate and γ is a small positive constant used for normalization. If soft competition is used, a Gaussian weighting function centered at the winning model is applied to all competing models. Every model is then updated proportional to the weight assigned to that model by this Gaussian weighting function such that

$$\mathbf{w}_i(n+1) = \mathbf{w}_i(n) + \eta \, \frac{\Lambda_{i,j}(n)e_i(n)\mathbf{x}(n)}{\gamma + \|\mathbf{x}(n)\|^2}, i = 1,\ldots,M,$$ (3.59)

where $\mathbf{w}_i$ is the weight vector of the ith model. Assuming the jth model wins competition, $\Lambda_{i,j}(n)$ is the weighting function defined by

$$\Lambda_{i,j}(n) = \exp\left(-\frac{\mathrm{d}_{ij}^2}{2\sigma^2(n)}\right),$$ (3.60)

where d_{ij} is the Euclidean distance between indices i and j, which is equal to $|j - i|$, $\eta(n)$ is the annealed learning rate, and $\sigma^2(n)$ is the Gaussian kernel width decreasing exponentially as n increases. The learning rate also exponentially decreases with n.

Soft competition preserves the topology of the input space, updating the models neighboring the winner; thus, it is expected to result in smoother transitions between models specializing in topologically neighboring regions (of the state space). However, the empirical comparison using BMI data between hard and soft competition update rules shows no significant difference in terms of model performance (possibly because of the nature of the data set). Therefore, we prefer to utilize hard competition rule because of its simplicity.

With the competitive training procedure, each model can specialize in local regions in the joint space. Figure 3.18 demonstrates the specialization of 10 trained models by plotting their outputs (black dots) with the common input data (40 sec long) in the 3D hand trajectory space. Each model's outputs are simultaneously plotted on top of the actual hand trajectory (red lines) synchronized with the common input. This figure shows that the input–output mappings learned by each model display some degree of localization, although overlaps are still present. These overlaps may be consistent with a neuronal multiplexing effect as depicted in Carmena et al. [26], which suggests that the same neurons modulate for more than one motor parameter (the x and y coordinates of hand position, velocity, and griping force).

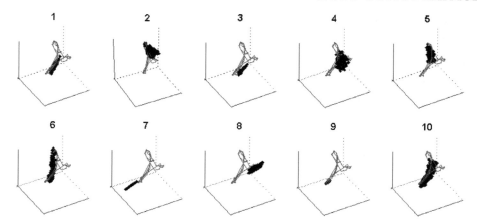

FIGURE 3.18: Demonstration of the localization of competitive linear models.

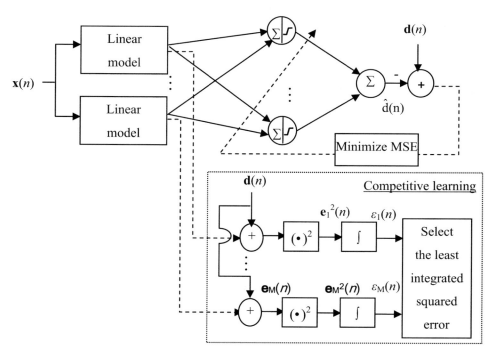

FIGURE 3.19: An overall diagram of the nonlinear mixture of competitive linear models.

The competitive local linear models, however, require additional information for switching when applied to BMIs because the desired signal that is necessary to select a winning model is not available in practice after training. A gate function as in the mixture of experts [47] utilizing input signals needs to be trained to select a local model. Here, we opt for an MLP that directly combines the predictions of all models. Therefore, the overall architecture can be conceived as a nonlinear mixture of competitive linear models [30] and is architecturally a TDNN. This procedure facilitates training of each model compared with the TDNN because only one linear model is trained at a time in the first stage, whereas only a relatively small number of weights are trained by error backpropagation [2] in the second stage.

The overall architecture of nonlinear mixture of competitive linear models is identical to a single hidden-layer TDNN as shown in Figure 3.19. However, the training procedure undertaken here is significantly different. The topology allows a two-stage training procedure that can be performed sequentially in offline training: first, competitive learning for the local linear models and then error backpropagation learning for the MLP. It is important to note that in this scheme, both the linear models and the MLP are trained to approximate the same desired response, which is the hand trajectory.

3.3   SUMMARY

Here, we have presented a comprehensive overview of input–output modeling techniques for BMIs. We would like to address here a few aspects of the modeling techniques that can influence performance. First, the tradeoffs between timing resolution (bin size) and memory depth needed to optimally solve the system identification problem for all models has not been directly addressed [57] and can present as a source of variability in model performance. Second, all of the variations in modeling (linear vs. nonlinear, feedfoward vs. recurrent, global vs. local) and enhancements shown improve the original solution in a marginal but statistically significant way. In more traditional signal-processing applications of these techniques, the differences in modeling performance per topology are often greater. This observation leads one to suspect either that all of the topologies are not optimal for BMI signal processing or there is some common property to the data (for rats, monkeys, and humans) that is influencing all of the results. For all BMI data in the literature, the reported performance ranges in general from 60% to 90% correlation of the model output with the true behavioral trajectories. However, because each research group is using their own data and the performance changes from task to task, animal to animal, and even from day to day, it is very difficult to compare results. A third aspect that complicates the comparison of decoding techniques is the simplicity of the movement tasks studied. The next generation of decoding models will have to assess performance in multipostural reaching, grasping, and holding under the influence of force.

Here, the possible advantages of nonlinear models may be better highlighted. All in all, the viability of input–output decoding the intent of motion contained in multidimensional spike trains collected from the motor cortex has been established by numerous research groups and should be considered an acquired scientific fact.

REFERENCES

1. Soderstrom, T., and P. Stoica, System Identification. 1989, New York: Prentice-Hall.
2. Haykin, S., Adaptive Filter Theory. 3rd ed. 1996, Upper Saddle River, NJ: Prentice-Hall International.
3. Ljung, L. *Black-box models from input–output measurements*, in IEEE Instrumentation and Measurement Technology Conference. 2001. Budapest, Hungary. doi:10.1109/IMTC.2001.928802
4. Kandel, E.R., J.H. Schwartz, and T.M. Jessell, eds. Principles of Neural Science. 4th ed. 2000, New York: McGraw-Hill.
5. Georgopoulos, A.P., A.B. Schwartz, and R.E. Kettner, *Neuronal population coding of movement direction.* Science, 1986. **233**(4771): pp. 1416–1419.
6. Georgopoulos, A., et al., *On the relations between the direction of two-dimensional arm movements and cell discharge in primate motor cortex.* Journal of Neuroscience, 1982. **2**: pp. 1527–1537.
7. Flament, D., and J. Hore, *Relations of motor cortex neural discharge to kinematics of passive and active elbow movements in the monkey.* Journal of Neurophysiology, 1988. **60**(4): pp. 1268–1284.
8. Wessberg, J., et al., *Real-time prediction of hand trajectory by ensembles of cortical neurons in primates.* Nature, 2000. **408**(6810): pp. 361–365.
9. Taylor, D.M., S.I.H. Tillery, and A.B. Schwartz, *Direct cortical control of 3D neuroprosthetic devices.* Science, 2002. **296**(5574): pp. 1829–1832. doi:10.1126/science.1070291
10. Serruya, M.D., et al., *Brain–machine interface: Instant neural control of a movement signal.* Nature, 2002. **416**: pp. 141–142. doi:10.1038/416141a
11. Sanchez, J.C., et al. *A comparison between nonlinear mappings and linear state estimation to model the relation from motor cortical neuronal firing to hand movements*, in SAB Workshop on Motor Control in Humans and Robots: on the Interplay of Real Brains and Artificial Devices. 2002. University of Edinburgh, Scotland.
12. Sanchez, J.C., et al. *Input–output mapping performance of linear and nonlinear models for estimating hand trajectories from cortical neuronal firing patterns*, in International Work on Neural Networks for Signal Processing. 2002. Martigny, Switzerland. doi:10.1109/NNSP.2002.1030025
13. Moran, D.W., and A.B. Schwartz, *Motor cortical representation of speed and direction during reaching.* Journal of Neurophysiology, 1999. **82**(5): pp. 2676–2692.

14. Kalaska, J.F., et al., *A comparison of movement direction-related versus load direction-related activity in primate motor cortex, using a two-dimensional reaching task.* Journal of Neuroscience, 1989. **9**(6): pp. 2080–2102.

15. Georgopoulos, A.P., et al., *Mental rotation of the neuronal population vector.* Science, 1989. **243**(4888): pp. 234–236.

16. Thach, W.T., *Correlation of neural discharge with pattern and force of muscular activity, joint position, and direction of intended next movement in motor cortex and cerebellum.* Journal of Neurophysiology, 1978. **41**: pp. 654–676.

17. Scott, S.H., and J.F. Kalaska, *Changes in motor cortex activity during reaching movements with similar hand paths but different arm postures.* Journal of Neurophysiology, 1995. **73**(6): pp. 2563–2567.

18. Todorov, E., *Direct cortical control of muscle activation in voluntary arm movements: A model.* Nature Neuroscience, 2000. **3**(4): pp. 391–398.

19. Gao, Y., et al. *A quantitative comparison of linear and non-linear models of motor cortical activity for the encoding and decoding of arm motions*, in the 1st International IEEE EMBS Conference on Neural Engineering. 2003. Capri, Italy.

20. Wu, W., et al. *Inferring hand motion from multi-cell recordings in motor cortex using a Kalman filter*, in SAB Workshop on Motor Control in Humans and Robots: On the Interplay of Real Brains and Artificial Devices. 2002. University of Edinburgh, Scotland.

21. Kalman, R.E., *A new approach to linear filtering and prediction problems.* Transactions of the ASME Journal of Basic Engineering, 1960. **82**(Series D): pp. 35–45.

22. Wiener, N., Extrapolation, Interpolation, and Smoothing of Stationary Time Series with Engineering Applications. 1949, Cambridge, MA: MIT Press.

23. Haykin, S., Neural Networks: A Comprehensive Foundation. 1994, New York: Macmillan; Toronto, Canada: Maxwell Macmillan.

24. Orr, G., and K.-R. Müller, Neural Networks: Tricks of the Trade. Vol. 1524. 1998, Berlin: Springer.

25. Chapin, J.K., et al., *Real-time control of a robot arm using simultaneously recorded neurons in the motor cortex.* Nature Neuroscience, 1999. **2**(7): pp. 664–670.

26. Carmena, J.M., et al., *Learning to control a brain–machine interface for reaching and grasping by primates.* PLoS Biology, 2003. **1**: pp. 1–16. doi:10.1371/journal.pbio.0000042

27. Shenoy, K.V., et al., *Neural prosthetic control signals from plan activity.* NeuroReport, 2003. **14**: pp. 591–597. doi:10.1097/00001756-200303240-00013

28. Horn, R.A., and C.R. Johnson, Topics in Matrix Analysis. 1991, New York: Cambridge University Press.

29. Shephard, N., *Maximum likelihood estimation of regression models with stochastic trend components.* Journal of the American Statistical Association, 1993. **88**(422): pp. 590–595. doi:10.2307/2290340

30. Kim, S.P., et al., *Divide-and-conquer approach for brain machine interfaces: Nonlinear mixture of competitive linear models.* Neural Networks, 2003. **16**(5–6): pp. 865–871. doi:10.1016/S0893-6080(03)00108-4

31. Hoerl, A.E., and R.W. Kennard, *Ridge regression: Biased estimation for nonorthogonal problems.* Technometrics, 1970. **12**(3): pp. 55–67.

32. Widrow, B., and S.D. Stearns, Adaptive Signal Processing. Prentice-Hall Signal Processing Series. 1985, Englewood Cliffs, NJ: Prentice-Hall.

33. Príncipe, J.C., N.R. Euliano, and W.C. Lefebvre, Neural and Adaptive Systems: Fundamentals Through Simulations. 2000, New York: Wiley.

34. Sanchez, J.C., et al. *Interpreting neural activity through linear and nonlinear models for brain machine interfaces*, in International Conference of Engineering in Medicine and Biology Society. 2003. Cancun, Mexico. doi:10.1109/IEMBS.2003.1280168

35. Rao, Y.N., et al. *Learning mappings in brain–machine interfaces with echo state networks*, in International Joint Conference on Neural Networks. 2004. Budapest, Hungary. doi:10.1109/ICASSP.2005.1416283

36. Sandberg, I.W., and L. Xu, *Uniform approximation of multidimensional myopic maps.* IEEE Transactions on Circuits and Systems, 1997. **44**: pp. 477–485.

37. Todorov, E., *On the role of primary motor cortex in arm movement control*, in Progress in Motor Control III, M. Latash, and M. Levin, eds. 2003, Urbana, IL: Human Kinetics.

38. Puskorius, G.V., et al., *Dynamic neural network methods applied to on-vehicle idle speed control.* Proceedings of the IEEE, 1996. **84**(10): pp. 1407–1420. doi:10.1109/5.537107

39. Werbos, P.J., *Backpropagation through time: What it does and how to do it.* Proceedings of the IEEE, 1990. **78**(10): pp. 1550–1560. doi:10.1109/5.58337

40. Lefebvre, W.C., et al., NeuroSolutions. 1994, Gainesville, FL: NeuroDimension.

41. Vapnik, V., The Nature of Statistical Learning Theory. Statistics for Engineering and Information Science. 1999, New York: Springer-Verlag. 304.

42. Jaeger, H., *The "Echo State" Approach to Analyzing and Training Recurrent Neural Networks*, GMD Report 148. 2001, Sankt Augustin, Germany: GMD-German National Research Institute for Computer Science.

43. Maas, W., T. Natschläger, and H. Markram, *Real-time computing without stable states: A new framework for neural computation based on perturbations.* Neural Computation, 2002. **14**(11): pp. 2531–2560. doi:10.1162/089976602760407955

44. Principe, J.C., B. De Vries, and P.G. Oliveira, *The gamma filter—A new class of adaptive IIR filters with restricted feedback.* IEEE Transactions on Signal Processing, 1993. **41**(2): pp. 649–656. doi:10.1109/78.193206

45. Ozturk, M.C., D. Xu, and J.C. Principe, *Analysis and Design of Echo State Network for Function Approximation*. Neural Computation, 2006. **19**: pp. 111–138.

46. Farmer, J.D., and J.J. Sidorowich, *Predicting chaotic time series*. Physical Review Letters, 1987. **50**: pp. 845–848. doi:10.1103/PhysRevLett.59.845

47. Jacobs, R., et al., *Adaptive mixture of local experts*. Neural Computation, 1991. **3**: pp. 79–87.

48. Fancourt, C., and J.C. Principe. *Temporal self-organization through competitive prediction*, in ICASSP. 1996. doi:10.1109/ICASSP.1997.595505

49. Cho, J., et al., *Self-organizing maps with dynamic learning for signal reconstruction*. Neural Networks, 2007. **20**(2): pp. 274–284. doi:10.1016/j.neunet.2006.12.002

50. Jordan, M.I., and R.A. Jacobs, *Hierarchical mixtures of experts and the EM algorithm*. Neural Computation, 1994. **6**: pp. 181–214. doi:10.1109/IJCNN.1993.716791

51. Weigend, A.S., M. Mangeas, and A.N. Srivastava, *Nonlinear gated experts for time-series— Discovering regimes and avoiding overfitting*. International Journal of Neural Systems, 1995. **6**(4): pp. 3773–399.

52. Zeevi, A.J., T. Meir, and V. Maiorov, *Error bounds for functional approximation and estimation using mixtures of experts*. IEEE Transactions on Information Theory, May 1998. **44**: pp. 1010–1025.

53. Jacobs, R.A., et al., *Adaptive mixtures of local experts*. Neural Computation, 1991. **3**: pp. 1–12. doi:10.1109/78.726819

54. Fancourt, C., and J.C. Principe, *Competitive principal component analysis for locally stationary time series*. IEEE Transactions on Signal Processing, 1998. **46**(11): pp. 3068–3081.

55. Pawelzik, K., J. Kohlmorgen, and K.-R. Muller, *Annealed competition of experts for a segmentation and classification of switching dynamics*. Neural Computation, 1996. **8**(2): pp. 340–356. doi:10.1007/BF00317973

56. Kohonen, T., *Analysis of a simple self-organizing process*. Biological Cybernetics, 1982. **44**(2): pp. 135–140. doi:10.1162/089976606774841585

57. Wu, W., et al., *Bayesian population decoding of motor cortical activity using a Kalman filter*. Neural Computation, 2005. **18**: pp. 80–118.

· · · ·

CHAPTER 4

Regularization Techniques for BMI Models

Additional Contributor: Sung-Phil Kim

In Chapter 3, we have demonstrated the design of linear and nonlinear filters which can be adapted for BMI applications. Despite the intrinsic sophistication of the BMI system, early tests with the simple linear filter (which merely combines the weighted bin count inputs) showed reasonable estimation of hand position (HP) from neuronal modulation. Therefore, it makes sense to fine-tune these linear models by means of advanced learning techniques. There are three major issues in the design of BMIs: the large number of parameters, irrelevant inputs, and the noise in the data due to spike sorting and nonstationarities. They all affect model performance in different ways. The first two can be handled using sophisticated signal processing techniques during offline analysis of the data. The latter issue of spike sorting is usually handled during the time of data collection, and is considered as a preprocessing step. Here we assume that the data have been appropriately spike sorted, and will focus on the signal processing techniques used in offline analysis.

When training BMI models the first challenge one encounters is how to deal with model overfitting when hundreds of neurons are used as model inputs. In the examples, we showed that the introduction of extra degrees of freedom not related to the mapping can result in poor generalization, especially in topologies where tap-delay memory structures are implemented in the neural input layer (i.e., the TDNN topology). The problem occurs with each additional memory delay element that scales the number of free parameters by the number of input neurons. This explosion in the number of free parameters also puts a computational burden on computing an optimal solution especially when the goal is to implement the BMI in low-power, portable hardware. A table of the number of free parameters for the topologies is described in Chapter 3.

The generalization of the model can be explained in terms of the bias-variance dilemma of machine learning [1], which is related to the number of free parameters of a model. The MIMO structure of BMIs built for the neuronal data presented here can have as few as several hundred to as many as several thousand free parameters. On one extreme, if the model does not contain enough parameters, there are too few degrees of freedom to fit the to be estimated, which results in bias errors. On the other extreme, models with too many degrees of freedom tend to overfit the function to be estimated. In terms of BMIs, models tend to err on the latter because of the large dimensionality of the input.

The first approach we presented for reducing the number of free parameters involved modifications to the model topology itself. More parsimonious models such as the recurrent multilayer perceptron (RMLP) [2–4] were studied. The topology of this model significantly reduced the number of free parameters by implementing feedback memory structures in hidden network layers instead of in the input where the dimensionality is large.

Besides modifying the model topology, a second statistical approach can be taken during model training. Regularization techniques [5] attempt to reduce 4the value of unimportant weights to zero, and effectively prune the size of the model topology. Regularized least squares (LS), subspace projections using partial least squares (PLS), or special memory structures can be used to reduce the number of free parameters [6]. These approaches are strictly statistical, and require lots of data and computation, are not trivial to use, and do not necessarily provide information about the importance of neurons. As an alternative, the number of inputs given to the models could be manually pruned using neurophysiological analysis of the correlation of neuronal function with behavior; however, it is difficult to know how it will affect BMI model performance. To overcome this issue, neural selection has also been attempted using sensitivity analysis and variable selection procedures [4, 7].

This chapter presents our efforts to deal with these problems which are basically divided between regularization of the optimal solution and channel selection. Moreover, once we have a well-tuned model, then the model itself can help us learn about the neurophysiology of the motor cortex. We also show the quantification of neural responses that is possible by "looking inside" the trained models.

4.1   LEAST SQUARES AND REGULARIZATION THEORY

The nature of the neural systems used in BMI model architectures creates MIMO systems with a large input space. Even with the LS solution, an issue of well-posedness follows immediately because the algorithm is formulated in very high dimensional spaces with finite training data available. The concept of well-posedness was proposed by Hadamard [8]. Regularization as a remedy for ill-posedness became widely known because of the work of Tikhonov [9], and also from Bayesian Learning for Neural Networks [10]. In solving LS problems, the Tikhonov regularization is essentially a trade-off between fitting training data and reducing solution norms. Consequently, it reduces the sensitivity of the solution to small changes in the training data and imposes stability on the ill-posed problem. Moreover, the significance of well-posedness related to generalization ability has been also revealed recently in statistical learning theory [11].

Recall that in LS, the multiple linear regression model is hypothesized as

$$y_n = W^\circ(\Phi_n) + v(n) \tag{4.1}$$

where W represents the model weights, Φ_n is the identity or a nonlinear function depending on whether the model is linear or nonlinear, and $v(n)$ is the modeling uncertainty. We will use Φ_n to mean $\Phi(x_n)$, where x is the neuronal input.

Denote the transformed data matrix as $D_W^T = [W(x_1), W(x_2), ..., W(x_N)]$, the correlation matrix by

$$R_W = D_W^V D_W / N \tag{4.2}$$

the cross-correlation vector $p_\Phi = \sum_{i=1}^{N} \Phi(x_i) y_i / N = D_\Phi^T \bar{y} / N$, and the Gram matrix

$$G_\Phi = D_\Phi D_\Phi^T = [\kappa(x_i, x_j)]_{N \times N} \tag{4.3}$$

Then, the LS solution is known to satisfy the following normal equations [8]

$$R_\Phi \hat{W} = p_\Phi \tag{4.4}$$

Unfortunately, the LS problem is not always well posed. According to Hadamard [8], a problem is well posed if the solution exists, is unique, and smoothly depends on data and parameters. More specifically, the following theorems demonstrate how to get a well-posed LS algorithm.

Theorem 4.1: *(uniqueness theorem of LS [12]) The least-squares estimate $\hat{W}$ is unique if and only if the correlation matrix R_Φ is nonsingular and the solution is given by*

$$\hat{W} = R_\Phi^{-1} p_\Phi \tag{4.5}$$

Theorem 4.2: *(pseudo-inverse [12]) If R_Φ is singular, there are infinitely many solutions to (4.5). Of them, there is a unique minimum-norm solution given by*

$$\hat{W} = D_\Phi^+ \bar{y} \tag{4.6}$$

Here, D_Φ^+ is the general pseudo-inverse of D_Φ given by

$$D_\Phi^+ = p \begin{bmatrix} S^{-1} & 0 \\ 0 & 0 \end{bmatrix} Q^T \tag{4.7}$$

and matrices P, Q, and S are given by the singular value decomposition (SVD) of D_Φ [13]

$$D_\Phi = Q \begin{bmatrix} S & 0 \\ 0 & 0 \end{bmatrix} P^{\mathrm{T}} \tag{4.8}$$

where P, Q are orthogonal matrices, and $S = \mathrm{diag}\{s_1, s_2, ..., s_r\}$ assuming the rank of D_Φ is r. s_i denotes the singular values of D_Φ and assumed $s_1 \geq s_2 \geq \cdots \geq s_r > 0$.

Remarks: The solution (4.6) includes (4.5) as special case, and together can be expressed as

$$\hat{W} = R_\Phi^+ p_\Phi \tag{4.9}$$

In an ideal world, the designer's job is finished with (4.9), but in the real world more caution should be taken. The fundamental issues are the finite precision of our computers and the noise—universal thermal noise, measurement errors, modeling uncertainty, etc. For example, when the modeling uncertainty $v(n)$ in (4.1) is white with zero mean and constant variance σ^2, the covariance matrix of the LS estimate (4.6) is [14]

$$\mathrm{cov}[\hat{W}] = \sigma^2 P \begin{bmatrix} S^{-2} & 0 \\ 0 & 0 \end{bmatrix} P^{\mathrm{T}} \tag{4.10}$$

If some singular values are too small, the variance of the estimate is pessimistically large. On the other hand, too small singular values also cause catastrophic numerical errors because of the finite precision [13]. The typical symptoms of this type of ill-posedness are large norm solutions and extreme sensitivity to small changes in training data.

The Tikhonov regularization is widely used to address this problem. Because small norm solutions are desirable, a regularization term is introduced in the LS cost function which penalizes the solution norm

$$\min_W R_{\mathrm{reg}} \left[W \in H_2, Z^N \right] = \left\| \bar{y} - D_\Phi W \right\|^2 + \lambda \left\| W \right\|_F^2 \tag{4.11}$$

In the Bayesian interpretation, the error square term is the likelihood and the regularized term is some priori knowledge on the norm of the solution, which corresponds to a Gaussian prior for the L_2 norm [14].

Theorem 4.3: *(Tikhonov regularization [9]) The Tikhonov regularized LS solution is*

$$\hat{W}_{\mathrm{TR}} = (D_\Phi^T D_\Phi + \lambda I)^{-1} D_\Phi^T \bar{y} \tag{4.12}$$

Remarks: More insights can be obtained through the SVD,

$$\hat{W}_{\mathrm{TR}} = P\mathrm{diag}\left(\frac{s_1}{s_1^2 + \lambda}, ..., \frac{s_r}{s_r^2 + \lambda}, 0, ..., 0\right)Q^{\mathrm{T}}\bar{y} \qquad (4.13)$$

Comparing it with the pseudo-inverse solution

$$\hat{W} = P\mathrm{diag}(s_1^{-1}, ..., s_r^{-1}, 0, ..., 0)Q^{\mathrm{T}}\bar{y} \qquad (4.14)$$

we see that the Tikhonov regularization modifies the singular values through the following regularization function:

$$H_{\mathrm{TR}}(x) = x^2 /(x^2 + \lambda) \qquad (4.15)$$

Note that $H_{\mathrm{TR}}(x) \to 1$ when x is large and $H_{\mathrm{TR}}(x) \to 1$ when x is small. In this sense, the Tikhonov regularization smoothly filters out the singular components that are small (relative to λ). This viewpoint actually brings new understanding into all kinds of regularization techniques as will be described in this chapter.

4.1.1   Ridge Regression and Weight Decay

When building linear models for BMI with neuronal activity that is highly variable with a large dynamic range of bin counts, the condition number of an input correlation matrix may be relatively large as shown in Chapter 3. In this situation, the performance of the model for reconstructing the movement trajectories from neural activity is highly variable because of the extra degrees of freedom that are not constrained. To reduce the condition number, we can, according to (4.12), add an identity matrix multiplied by a white noise variance to the correlation matrix, which is known as ridge regression (RR) in statistics [15] when LS are being solved as in the case of the Wiener solution

We can also use a similar regularizer in the iterative updates based on the least mean squares (LMS), or normalized LMS (NLMS) algorithm, which is called weight decay [16]. Rewriting (4.11), the criterion function of the regularized solution is

$$J(\mathbf{w}) = E[\|\mathbf{e}\|^2] + \delta\|\mathbf{w}\|^2 \qquad (4.16)$$

where the additional term $\delta\|\mathbf{w}\|^2$ smoothes the cost function. The instantaneous gradient that corresponds to this equation can be written as

$$\mathbf{w}(n+1) = \mathbf{w}(n) + \eta\hat{\nabla}\zeta(n) - \delta\mathbf{w}(n) \qquad (4.17)$$

where $\hat{\nabla}\zeta(n) = \partial E\left[\|\mathbf{e}(n)\|^2\right]/\partial\mathbf{w}(n)$, which is called the weight decay gradient update. Both RR and weight decay can be viewed as implementations of a Bayesian approach to complexity control in supervised learning using a zero-mean Gaussian prior [10].

The choice of the amount of regularization (δ) plays an important role in the generalization performance, because there is a trade-off between the condition number and the achievable MSE for a particular δ. A larger δ can decrease the condition number at the expense of increasing the MSE, whereas a smaller δ can decrease the MSE but also increase the condition number. Larsen et al. [17] proposed that δ can be optimized by minimizing the generalization error with respect to δ. Following this procedure, we utilize the K-fold cross-validation [18], which divides the data into K randomly chosen disjoint sets, to estimate the average generalization error empirically,

$$\hat{\xi} = \frac{1}{K}\sum_{k=1}^{K}\varepsilon_k \qquad (4.18)$$

where ε_k is MSE of the validation for the kth set. Then, the optimal regularization parameter is learned by using gradient descent,

$$\delta(n+1) = \delta(n) - \eta\frac{\partial\hat{\xi}(n)}{\partial\delta} \qquad (4.19)$$

where $\hat{\xi}(n)$ is an estimate computed with $\delta(n)$, and $\eta > 0$ is a learning rate. See Reference [17] for the procedure of estimation of $\partial\hat{\xi}(n)/\partial\delta$ using weight decay. Once the procedure is applied, the following change in weight distribution can be obtained as in Figure 4.1. Here, a linear BMI was trained with both weight decay and standard NLMS. As we can see, the solid line has many more weights with the value of zero. This means that during adaptation, the extra degrees of freedom in the model were effectively eliminated. The elimination of the influence of many weights resulted in an increase in the testing correlation coefficient, which also had a corresponding decrease in variance as shown in Table 4.1.

4.1.2   Gamma Filter

The large number of parameters in decoding models is caused not only by the number of neurons but by the number of time delays required to capture the history of the neuron firings over time. This attribute of the neural input topology was especially evident with the analysis of the TDNN. This problem is compounded in the context of neural decoding we showed that the time resolution of the neural rate coding can influence performance. If smaller time bins are used in decoding, additional numbers of parameters could be introduced. For example, although we use a 10-tap delay line for 100-msec bin sizes, the size of the delay line is variable depending on the bin size (e.g., if we use a 50-msec time bin, then the number of time lags doubles for an equivalent memory depth).

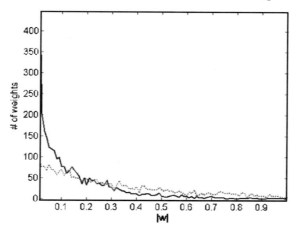

FIGURE 4.1: Histogram of the magnitudes of weights over all the coordinates of hand position (HP), trained by weight decay (solid line), and NMLS (dotted line).

A generalized feedforward filter (GFF) provides a signal processing framework to incorporate both finite (FIR) and infinite impulse response (IIR) characteristics into a single linear system by using a local feedback structure in the delay line as a generalized delay operator [19]. As shown in Figure 4.2, an input signal is delayed at each tap by a delay operator defined by specific transfer function $G(z)$. Note that when $G(z) = z^{-1}$, it becomes an FIR filter. The transfer function of an overall system $H(z)$ is stable when $G(z)$ is stable because

$$H(z) = \sum_{k=0}^{K} w_k \left(G(z) \right)^k \qquad (4.20)$$

where K is the number of taps. It has been shown that a GFF can provide trivial stability checks if $G(z)$ is a first- or second-order polynomial in z, and easy adaptation while decoupling the memory

TABLE 4.1: Comparison of testing performance for weight decay and NLMS		
	NLMS	**WEIGHT DECAY**
Testing a correlation coefficient for reaching task	0.75 ± 0.20	0.77 ± 0.18

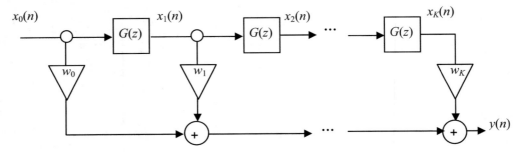

FIGURE 4.2: An overall diagram of a generalized feedforward filter [19]. $x_0(n)$ is an instantaneous input and $y(n)$ is a filter output.

depth from the filter order [19]. Memory depth in the present context is defined as the center of mass of the impulse response of the last tap of the GFF filter [19].

The gamma filter is a special case of the GFF with $G(z) = \mu/(z - (1 - \mu))$, where μ is a feedback parameter. The impulse response of the transfer function from an input to the kth tap, denoted as $g_k(n)$, is given by

$$g_k(n) = Z^{-1}(G_k(z)) = Z^{-1}\left(\left(\frac{\mu}{z-(1-\mu)}\right)^k\right) = \binom{n-1}{k-1}\mu^k(1-\mu)^{n-k}u(n-k) \qquad (4.21)$$

where $Z^{-1}(\cdot)$ indicates the inverse z transform and $u(n)$ the step function. When $\mu = 1$, the gamma filter becomes an FIR filter. The stability of the gamma filter in adaptation is guaranteed when $0 < \mu < 2$ because of a local feedback structure.

The memory depth D with a feedback parameter μ in the Kth-order gamma filter is given by

$$D = \frac{K}{\mu} \ \text{ for } \mu < 1, \text{ or } D = \frac{K}{2-\mu} \ \text{ for } \mu > 1. \qquad (4.22)$$

If we defined the resolution $R \equiv \mu$, the property of the gamma delay line can be described as

$$K = D \times R \ \text{ for } \mu < 1, \text{ or } K = D \times (2 - R) \ \text{ for } \mu > 1. \qquad (4.23)$$

This property shows that the gamma filter decouples the memory depth from the filter order by adjusting a feedback parameter (μ). In the case of $\mu = 1$ (i.e., the FIR filter), the resolution is maximized, whereas the memory depth is minimized for a given filter order. But this choice sometimes results in overfitting when a signal to be modeled requires more time delays than the number of descriptive parameters. Therefore, the gamma filter with the proper choice of a feedback parameter can avoid overfitting by reducing the number of parameters required to span a certain memory depth. Sandberg and Xu showed that the gamma neural model, which is obtained by substituting the tap delay line in the TDNN by a gamma memory, is still a universal mapper in functional spaces [20].

The tap weights can be either computed by LS or updated using NLMS, and therefore the computational complexity is of the same order of FIR filters. The feedback parameter μ can also be adapted from the data using an LMS type of update [21]. However, instead of adaptively finding μ, we can search the best combination of K and μ through cross-validation.

In Table 4.2, we can see that by using the Gamma architecture the number of model parameters can be reduced by 40%. The reduction yielded a 3% increase in the testing correlation coefficient with a reduction in the variance.

4.1.3 Subspace Projection

In addition to unnecessary weights in the model, one of the challenges in the design of decoding models for BMIs is that some neuron firings are not substantially modulated during task performance, and they only add noise to the solution. In the overall representation contained within these neural systems, the superfluous neurons may be coding other kinematics, sensory information, or are involved in internal signaling within the network. Although these neurons are important to the overall function of the network, they may not be correlated with the particular behavior

TABLE 4.2 Comparison of testing performance and number of parameters for the Gamma architecture and NLMS (1-sec memory depth)

	NLMS	GAMMA
Testing correlation coefficient for a reaching task	0.75 ± 0.20	0.78 ± 0.19
Number of parameters	2973	1191

studied during the BMI experiment.[1] In addition, some neuron firings are correlated with each other, which creates problems in the estimation of LS (R may become rank deficient); thus, it may be advantageous to blend these inputs to improve model performance. Subspace projection, which can at the same time reduce the noise, and blend correlated input signals together, may be very beneficial. Moreover, it also reduces the number of degrees of freedom in the multichannel data, and consequently decreases the variance of the model for the same training data size. Here, we introduce a hybrid subspace projection method that is derived by combining the criteria of PCA and PLS. Then, we will design the subspace Wiener filter based on this hybrid subspace projection for BMIs.

PCA, which preserves maximum variance in the subspace data, has been widely adopted as a projection method [14]. The projection vector $\boldsymbol{w}_{\mathrm{PCA}}$ is determined by maximizing the variance of the projection outputs as

$$\boldsymbol{w}_{\mathrm{PCA}} = \arg\max_{w} \quad J^{\mathrm{PCA}}(\boldsymbol{w}) = E\left[\left\|\boldsymbol{x}^{\mathrm{T}}\boldsymbol{w}\right\|^{2}\right] = \boldsymbol{w}^{\mathrm{T}}\mathbf{R}_{s}\boldsymbol{w} \tag{4.24}$$

where $\mathbf{R}_s$ is the input covariance matrix computed over the neuronal space only (it is an $M \times M$ matrix where M is the number of neurons). $\boldsymbol{x}$ is an $M \times 1$ instantaneous neuronal bin count vector. It has been well known that $\boldsymbol{w}_{\mathrm{PCA}}$ turns out to be the eigenvector of $\mathbf{R}_s$ corresponding to the largest eigenvalues. Then an $M \times S$ projection matrix that constructs an S-dimensional subspace consists of S eigenvectors corresponding to the S largest eigenvalues. However, PCA does not exploit information in the joint space of the input (neuronal) and desired (behavioral) response. This means that there may be directions with large variance that are not important to describe the correlation between input and desired response (e.g., some neuronal modulations related to the BMI patient's anticipation of reward might be substantial, but less useful for the direct estimation of movement parameters), but will be preserved by the PCA decomposition.

One of the subspace projection methods to construct the subspace in the joint space is PLS, which seeks the projection maximizing the cross-correlation between the projection outputs and desired response [22]. Given an input vector $\boldsymbol{x}$, and a desired response d, a projection vector of PLS, $\boldsymbol{w}_{\mathrm{PLS}}$, maximizes the following criterion,

$$\boldsymbol{w}_{\mathrm{PLS}} = \arg\max_{w} \quad J^{\mathrm{PLS}}(\boldsymbol{w}) = E\left[\left(\boldsymbol{x}^{\mathrm{T}}\mathbf{w}\right)d\right] = E\left[\boldsymbol{w}^{\mathrm{T}}(\boldsymbol{x}d)\right] = \boldsymbol{w}^{\mathrm{T}}\boldsymbol{p} \tag{4.25}$$

---

[1]At the time of surgery, the targeting of neurons is related to the broad functional specialization of each area of cortex. The ability to target specific neurons can be achieved through sensory, motor, or electrical stimulation. However, it is common to obtain the activity of neurons not related to the motor task of interest.

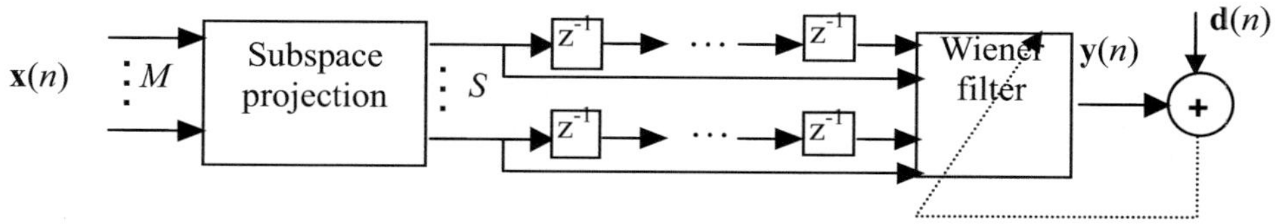

FIGURE 4.3: The overall diagram of the subspace Wiener filter. $y(n)$ denotes the estimated HP vector. There are $L - 1$ delay operators (z^{-1}) for each subspace channel.

where p is defined as an $M \times 1$ cross-correlation vector between x and d. The consecutive orthogonal PLS projection vectors are computed using the deflation method [14].

There have been efforts to find a better projection that can combine the properties of PCA and PLS. The continuum regression (CR), introduced by Stone and Brooks [23], attempted to blend the criteria of LS, PCA, and PLS. Recently, we proposed a hybrid criterion function similar to the CR, together with a stochastic learning algorithm to estimate the projection matrix [24]. The learned projection can be either PCA, PLS, or combination of the two. A hybrid criterion function combining PCA and PLS is given by

$$J(w,\lambda) = \frac{(w^{\mathrm{T}} p)^{2\lambda} (w^{\mathrm{T}} \mathbf{R_s} w)^{1-\lambda}}{w^{\mathrm{T}} w} \qquad (4.26)$$

where λ is a balancing factor between PCA and PLS. This criterion covers the continuous range between PLS ($\lambda = 1$) and PCA ($\lambda = 0$).[2] Because the log function is monotonically increasing, the criterion can be rewritten as,

$$\log(J(w,\lambda)) = \lambda \log\left(w^{\mathrm{T}} p\right)^2 + (1-\lambda)\log(w^{\mathrm{T}} \mathbf{R_s} w) - \log\left(w^{\mathrm{T}} w\right) \qquad (4.27)$$

We seek to maximize this criterion for $0 \le \lambda \le 1$. There are two learning algorithms derived in [24] to find w (one is based on gradient descent and the other is based on the fixed-point algorithm), but we opted to use the fixed-point learning algorithm here because to its fast convergence and independence of learning rate. The estimation of w at the $k+1$th iteration in the fixed-point algorithm is given by

$$w(k+1) = (1-T)w(k) + T\left[\frac{\lambda p}{w(k)^T p} + \frac{(1-\lambda)\mathbf{R_s} w(k)}{w(k)^T \mathbf{R_s} w(k)}\right] \qquad (4.28)$$

[2]The CR covers LS, PLS, and PCA. However, because we are only interested in the case when subspace projection is necessary, LS can be omitted in our criterion.

TABLE 4.3: Comparison of testing performance for PLS and NLMS

	NLMS	PLS
Testing correlation coefficient for a reaching task	0.75 ± 0.20	0.77 ± 0.18

with a random initial vector $w(0)$. $T (0 < T < 1)$ is a balancing parameter to remove the oscillating behavior near convergence. The convergence rate is affected by T that produces a trade-off between the convergence speed and the accuracy. Note that the fastest convergence can be obtained with $T = 1$. The consecutive projection vectors are also learned by the deflation method to form in each column of a projection matrix W. After projection onto the subspace by W, we embed the input signal at each channel with an L-tap delay line and design the Wiener filter to estimate the HP. Figure 4.3 illustrates the overall diagram of the subspace Wiener filter. For this architecture, the testing performance of the model also improved with a decrease in CC variance as shown in Table 4.3.

4.2   CHANNEL SELECTION

Because our ultimate goal for BMIs is to design the most accurate reconstructions of hand kinematics from cortical activity using adaptive signal processing techniques, it seems natural to equate here neural importance (selection of important neurons) to model fitting quality. Moreover, these measures should be compared with the available neurophysiological knowledge, with the hope that we can understand better the data, and enhance our methodologies and, ultimately, the performance of BMIs. Therefore, the importance of neurons will be ascertained using three techniques:

- Sensitivity analysis through trained linear and nonlinear models
- L_1-norm penalty pruning
- Real-time input selection

Given a set of data, we would like to evaluate how well these three methodologies are able to find important neurons for building BMI models. Second, we would like to use this information to tackle the model generalization issues encountered in BMIs. The goals of the study are formulated in the following questions:

- Can our methods automatically indicate important neurons for the prediction of the kinematic variables in the tasks studied?

- In this model-based framework, can better BMIs be built using a subset of important neurons?

It is well known that neurons vary in their involvement in a given task [25]. However, quantifying neuronal involvement for BMI applications is still an ongoing area of research. This is where BMI modeling is an asset, because once trained, the model implicitly contains the information of how cells contribute for the mapping. The difficulty is that the assessment is in principle dependent on the type of model chosen to predict the kinematic variables and its performance (model-dependent). We will first compare the linear FIR model and nonlinear RMLP. Our second question quantifies the change in performance when only a small subset of cells is used to build the BMI. In principle, one could think that any subset of cells will perform worse than the whole ensemble, but because of the poor generalization of large models, performance may in fact be better in a test set with a reduced number of important cells. Of course, this also makes BMIs more dependent on the stability over time of these cells, and in the long run we have shown that performance can either worsen or improve. The ultimate goal is to improve understanding of how cells encode kinematic parameters so that better "gray-box" models can be built using the underlying information in neural recordings.

Assumptions for ranking the importance of a neuron.   We would like to obtain an automatic measure of each cell's contribution to encoding motor parameters for a given task, which we call the cell importance. For this reason, a structured approach is taken to ascertaining the importance of neurons with the three methods described earlier. Our methodological choices, however, are not free from assumptions. First, the methods assume stationarity in the data. A snapshot of neural activity is taken and importance is ascertained without addressing time variability in the recordings, which is a shortcoming. Second, despite the highly interconnected nature of neural structures, importance is often computed independently for each individual neuron. With this independence assumption, it is difficult to quantify the importance of pairs, triples, etc., of cells. In contrast, the model sensitivity analysis considers covariations in firing rate among groups of cells in the neural ensemble, but depends on the type of model utilized. Third, a technique may only consider the instantaneous neural activity, whereas other techniques include memory structures (tap delay lines). Finally, each technique for ascertaining importance focuses on different neuronal firing features.

4.2.1   Sensitivity-Based Pruning

With the weights of the trained linear and nonlinear networks, we have a tool with which we can identify the neurons that affect the output the most. A sensitivity analysis, using the Jacobian of the output vector with respect to the input vector, tells how each neuron's spike counts affect the output given the data of the training set. Because the model topology can affect the interpretation of sensitivity, we will first examine sensitivities through the FIR filter that will serve as the "control model" throughout this book. The procedure for deriving the sensitivity for a feedforward topology is an application of the chain rule [26]. For the case of the FIR filter, differentiating the output with respect to the input [see (3.8)] directly yields a sensitivity with respect to each neuronal input i in (4.29).

$$\frac{\partial y_j}{\partial x_i} = \mathbf{W}_{10(i-1)+1:10(i-1)+10,j} \tag{4.29}$$

Hence, a neuron's importance can be determined by simply reading the corresponding weight value[3] in the trained model, if the input data for every channel is power-normalized. Because this is not the case for neural data, the neuron importance is estimated in the vector Wiener filter by multiplying the absolute value of a neuron's sensitivity with the standard deviation of its firing computed over the data set[4] as in (4.30). To obtain a scalar sensitivity value for each neuron, the weight values are also averaged over the 10 tap delays and three output dimensions.

$$\text{Sensitivity}_i = \sigma_i \frac{1}{2} \sum_{j=1}^{3} \frac{1}{10} \sum_{k=1}^{10} \left| \mathbf{W}_{10(i-1)=k,j} \right| \tag{4.30}$$

The procedure for deriving the sensitivity for a feedfoward multilayer perceptron (MLP), also discussed in [26], is again a simple application of the chain rule through the layers of the network topology as in (4.31):

$$\frac{\partial y_2(t)}{\partial x(t)} = \frac{\partial y_2(t)}{\partial y_1(t)} \frac{\partial y_1(t)}{\partial x(t)} \tag{4.31}$$

In the case of a nonlinear, dynamical system like the RMLP, the formulation must be modified to include time. Because the RMLP model displays dependencies over time that results from feedback in the hidden layer, we must modify this procedure [26]. Starting at each time t, we compute the sensitivities in (4.31) as well as the product of sensitivities clocked back in time. For example, using the RMLP feedforward equations [see (3.29) and (3.30)], we can compute at $t = 0$, the chain rule shown in (4.32). D_t is the derivative of the hidden layer nonlinearity evaluated at the operating point shown in (4.33). Notice that at $t = 0$ there are no dependencies on y_1. If we clock back one cycle, we must now include the dependencies introduced by the feedback, which is shown in (4.34). At each

---

[3]In this analysis, we consider the absolute values of the weights averaged over the output dimensions and the 10-tap delays per neuron.

[4]By multiplying the model weights by the firing standard deviation, we have modified the standard definition of sensitivity; however, for the remainder of this analysis we will refer to this quantity as the model sensitivity.

clock cycle back in time, we simply multiply an additional $\mathbf{W}_f^T D_{t-i}$. The general form of the sensitivity calculation is shown in (4.35). Experimentally, we determined that the effect of an input decays to zero over a window of 20 samples (Figure 4.4). At each time t, the sensitivity of the output with respect to the input is represented as the sum of the sensitivities over the 20-sample window.

$$\frac{\partial y_2(t)}{\partial x(t)} = \mathbf{W}_2^T D_t \mathbf{W}_1^T \tag{4.32}$$

$$\mathbf{D} = \begin{bmatrix} f'(z_1^1) & 0 & \cdots & 0 \\ 0 & f'(z_1^2) & \ddots & \vdots \\ \vdots & \ddots & \ddots & 0 \\ 0 & \cdots & 0 & f'(z_1^n) \end{bmatrix} \tag{4.33}$$

$$\frac{\partial y_2(t)}{\partial x(t-1)} = \mathbf{W}_2^T D_t \mathbf{W}_f^T D_{t-1} \mathbf{W}_1^T \tag{4.34}$$

$$\frac{\partial y_2(t)}{\partial x(t-\Delta)} = \mathbf{W}_2^T D_t \prod_{i=1}^{\Delta} \left(\mathbf{W}_f^T D_{t-i} \right) \mathbf{W}_1^T \tag{4.35}$$

Compared to the FIR model, which produces a static measure of sensitivity, the RMLP produces a time-varying sensitivity that we will now use to analyze three similar movements from

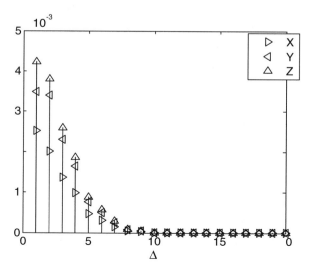

FIGURE 4.4: Sensitivity at time t for a typical neuron as a function of Δ.

the testing trajectory; specifically, reaching movements shown in the testing plots of Figure 4.5. Notice that movements are similar except that movement 3 does not have a decrease in the y coordinate. To visualize trends in how the input affects the output, we plot the neuronal activity (input) along with the computed sensitivity (averaged over 104 neurons and 20 delays) in Figure 4.5b–c. We take first a macroscopic view of both the neuronal activity and sensitivity by summing over 104 neurons at each time bin. Comparing Figure 4.5a with Figure 4.5b, we see that it is difficult to visually extract features that relate neuronal firing counts to hand movement. Despite this fact, a few trends arise:

- Neuronal activity consists of a time-varying firing rate around some mean firing rate.
- Movements 1 and 3 seem to show increased neuronal activity at the beginning and end of a movement, whereas 2 does not.
- All three movements contain a decrease in neuronal activity during the peak of the movement.

With these three trends in mind, we now include the sensitivity plot estimated for the RMLP in Figure 4.5c. We can observe the following:

- In general, the network becomes more sensitive during all three movements.
- Sensitivity is large when velocity is high.
- The sensitivity shows large, sharp values at the beginning and end of the movement, and a plateau during the peak of the movement.

From sensitivity peaks during the movements in Figure 4.5c, we ascertain that the sensitivity analysis is a requirement for relating neuronal activity to behavior. Without the model-based sensitivity, finding relationships in the raw data is difficult. Now that we have a mesoscopic view of how the ensemble of sampled neuronal activity affects the output of the network, we change our focus, and use the model-based sensitivity to "zoom in" on the important individual neurons. We are interested in learning why the individual neurons in a given cortical region are necessary for constructing the network output movement. Using the sensitivity analysis, we can select neurons that are most closely associated with the reaching movements (i.e., neurons that elicit large perturbations in the output with small perturbations in firing).

Neurons for a particular movement are selected for the RMLP by choosing the maximum of a sorted list of neuronal sensitivities computed by averaging the time-varying sensitivity in (4.35) over the interval of a movement and all three coordinate directions. For comparison, the neurons from the FIR filter are sorted directly using (4.30). The sorted ensemble neuronal sensitivities for

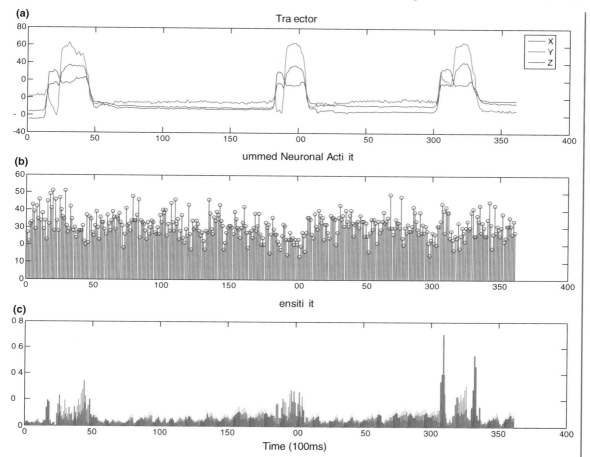

FIGURE 4.5: RMLP time-varying sensitivity. (a) X, Y, and Z desired trajectories for three similar movements. (b) Neuronal firing counts summed (at each time bin) over 104 neurons. (c) Sensitivity (averaged over 104 neurons) for three coordinate directions.

both the FIR and RMLP models (Figure 4.6) shows an initially sharp decrease from maximum indicating that only a few neurons are required for outlining the movement reconstruction. Of the 104 neurons from the animal used in this study, the 10 highest sensitivity neurons for a given movement are presented in the image plot in Figure 4.6, and are grayscale coded by cortical area (PP, posterior parietal; M1, primary motor; PMd, premotor dorsal; M1/PMd, ipsilateral). The highest sensitivity neurons are primarily distributed over the PP and M1. Of the 104 neurons, 7 of the 10 highest-ranking neurons are common for the WF and RMLP. By experimentally verifying the

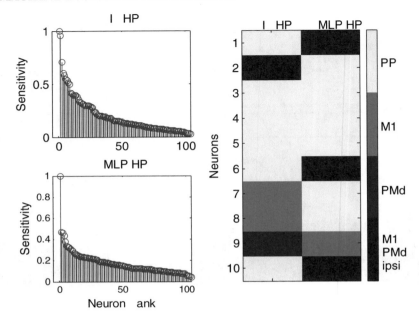

FIGURE 4.6: Reaching task neuronal sensitivities sorted from minimum to maximum for a movement. The 10 highest sensitivities are labeled with the corresponding neuron.

effect of computing the sensitivities through both WF and RMLP topologies, we have found that sensitivity-based selection of neurons is not heavily dependent on the model topology, even with two distinct topologies (linear-feedforward vs. nonlinear-feedback) are utilized [4].

As a further check of the sensitivity importance measure, we trained three identical RMLP networks of five hidden PEs, one with the 10 highest sensitivity neurons, one with the 84 intermediate sensitivity neurons, and one with the 10 lowest sensitivity neurons. A plot of the network outputs in Figure 4.7a, b, and c shows that the reduced set of highest sensitivity neurons does a good job of capturing the peaks of the movement, whereas the lowest sensitivity neurons are unable to discover the mapping from spike trains to hand movements. The remaining intermediate sensitivity neurons contain information about the movement but it is not enough to capture the full detail of the trajectory, although it does improve the overall trajectory fitting. Using the cumulative error metric (CEM)[5] for BMI performance in movements, we see that the curves approach the bisector of the space as sensitive neurons are dropped from the mapping, indicating a decrease in performance.

[5]CEM(r) is the estimated probability that the radius of the error vector is less than or equal to a certain value r.

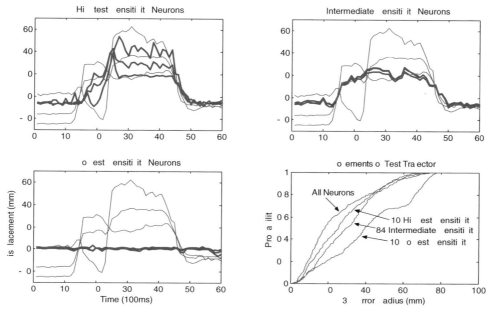

FIGURE 4.7: Testing outputs for RMLP models trained with subsets of neurons. (a–c) *x*, *y*, and *z* trajectories (bold) for one movement (light) from three RMLPs trained with the highest, intermediate, and lowest sensitivity neurons. (d) Cumulative error metric decreases as sensitive neurons are dropped.

This result verifies that the method of RMLP model-based sensitivity analysis for BMIs is able to produce a graded list of neurons involved in the reconstructing a hand trajectory.

We now need to address why the neurons with the highest sensitivity are important for the production of the output movement. The answer is quite simple and can be obtained by plotting the binned firing counts for highest and lowest sensitivity neurons in the vicinity of a reaching movement (Figure 4.8). In the top subplot of each column, we have included the first reaching movement as in the above simulations. For the highest sensitivity neurons, we can see some strong correlations between the neuronal firing and the rest/food, food/mouth, and mouth/rest movements. Neurons 4, 5, 7, 19, and 26 (PP) display increased firing during the rest/food movement, and are zero otherwise. Neurons 93 and 104 (M1) show a firing increase during the food/mouth movement. Neuron 45 (M1) does not correlate with the movement directly but seems to fire right before the movement, indicating that it may be involved in movement initiation. Neuron 84 (M1) fires only single pulses during the transitions of the movement. In the firing patterns of these neurons, we see the effect of the decay of the time-varying sensitivity at time *t*, which can be influenced by samples 2 sec in the past.

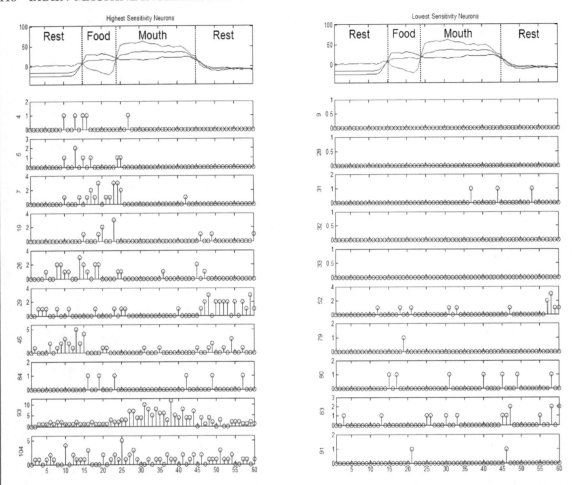

FIGURE 4.8: Neuronal firing counts from the 10 highest and lowest sensitivity neurons time-synchronized with the trajectory of one reaching movement.

All of the lowest sensitivity neurons primarily contain single firings, which do not display any measurable correlations to the movement. The data from these neurons are relatively sparse when compared to the highest sensitivity neurons, indicating that the network is not using them for the mapping.

Cellular Importance for Cursor Control Tasks.   We extend our analysis methodology for selecting important neurons for constructing the BMI mapping in a cursor control task. This time, however, the analysis will include all the kinematic variables available, HP, hand velocity (HV), and grip force (GF) from a rhesus monkey's neural recordings. For each variable, separate RMLPs were trained, and the sensitivity was computed as in (4.35). For the two-dimensional (2D) cursor task, sensitivities were averaged over the *x/y* coordinates and an arbitrary 3000 sample window because no clear marker of movement/nonmovement could be discriminated. The sorted position, velocity, and gripping force neuronal sensitivities for each session's ensemble of neurons is plotted in Figure 4.9a and b. Like the reaching task cellular analysis, an initially sharp decrease from maximum

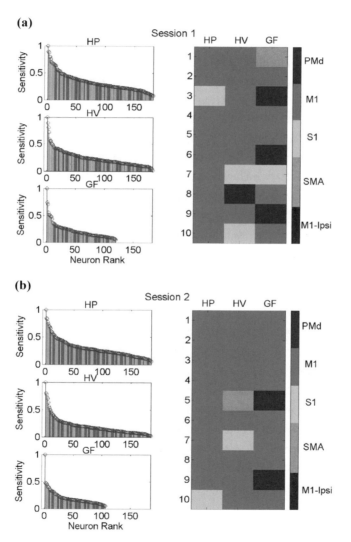

FIGURE 4.9: Sensitivity-based neuronal ranking for HP and HV for two recording sessions using an RMLP: (a) session 1 and (b) session 2. The cortical areas corresponding to the 10 highest ranking HP, HV, and GF neurons are given by the color map. HV, hand velocity; GF, grip force.

indicates that there are a few neurons that are significantly more sensitive than the rest of the ensemble. For HP and HV, the most important neurons are again mostly located in the primary motor cortex (M1). For the GF, we can also see that initially during first recording session multiple brain areas are contributing (30% premotor dorsal, 10% supplementary motor associative, 10% somatosensory), but many of these cells are replaced during session two with cells from M1.

4.2.2 L_1-norm Penalty Pruning

Recent studies on statistical learning have revealed that the L_1-norm penalty sometimes provides a better solution as a regularization method in many applications than the L_2-norm penalty [27]. Basically, the L_1-norm based regularization methods can select input variables more correlated with outputs as well as provide sparser models, compared to the L_2-norm based methods. Linear Absolute Shrinkage and Selection Operator (LASSO) has been a prominent algorithm [28] among the L_1-norm based regularization methods. However, its implementation is computationally complex. Least Angle Regression (LAR), which was recently proposed by Efron et al., provides a framework to incorporate LASSO and forward stagewise selection [29]. With LAR, the computational complexity in the learning algorithm can be significantly reduced.

The LAR algorithm has been recently developed to accelerate computation and improve performance of forward model selection methods. It has been shown in Efron et al. that simple modifications to LAR can implement the LASSO and the forward stagewise linear regression [29]. Essentially, the LAR algorithm requires the same order of computational complexity as ordinary least squares (OLS).

The selection property of LAR, which leads to zeroing model coefficients, is preferable for sparse systems identification when compared to regularization methods based on the L_2-norm penalty. Also, the analysis of the selection process often provides better insights into the unknown system than the L_2-norm based shrinkage methods.

The LAR procedure starts with an all-zero coefficients initial condition. The input variable having the most correlation with desired response is selected. We proceed in the direction of the selected input with a step size which is determined such that some other input variable attains as much correlation with the current residual as the first input. Next, we move in the equiangular direction between these two inputs until the third input has the same correlation. This procedure is repeated until either all input variables join the selection, or the sum of coefficients meets a preset threshold (constraint). Note that the maximum correlation between inputs and the residual decreases over successive selection step in order to decorrelate the residual with inputs. Table 4.4 summarizes the details of the LAR procedure [29].

The illustration in Figure 4.10 (taken from Efron et al. [29]) will help to explain how the LAR algorithm proceeds. In this figure, we start to move on the first selected input variable x_1 until the next variable (x_2 in this example) has the same correlation with the residual generated by x_1. μ_1 is the unit

TABLE 4.4: Procedure of the LAR algorithm

Given an $N \times M$ input matrix $\mathbf{X}$ (each row being M-dimensional sample vector), and an $N \times 1$ desired response matrix $\mathbf{Y}$, initialize the model coefficient $\boldsymbol{\beta}_i = 0$, for $i = 1,\dots, M$, and let $\boldsymbol{\beta} = [\boldsymbol{\beta}_1, \dots, \boldsymbol{\beta}_M]^T$

Then the initial LAR estimate becomes $\hat{\mathbf{Y}} = \mathbf{X}\boldsymbol{\beta} = 0$.

Transform $\mathbf{X}$ and $\mathbf{Y}$ such that $\dfrac{1}{N}\sum_{i=1}^{N} x_{ij} = 0$, $\dfrac{1}{N}\sum_{i=1}^{N} x_{ij}^2 = 1$, $\dfrac{1}{N}\sum_{i=1}^{N} y_i = 0$, for $j = 1,\dots, M$.

a) Compute the current correlation $c = \mathbf{X}^T (\mathbf{Y} - \hat{\mathbf{Y}})$

b) Find $C_{\max} = \max_j \left\{ |c_j| \right\}$, and a set A = $[j: |c_j| = C_{\max}]$.

c) Let $\mathbf{X}_a = [\dots, \text{sign}(c_j)\mathbf{x}_j, \dots]$ for $j \in$ A.

d) Let $\boldsymbol{\Phi} = \mathbf{X}_a^T \mathbf{X}_a$, and $\alpha = (\mathbf{1}_a^T \boldsymbol{\Phi}^{-1} \mathbf{1}_a)^{-1}$, where $\mathbf{1}_a$ is a vector of 1's with a length equal to size of A.

e) Compute the equiangular vector $\mu = \mathbf{X}_a(\alpha \boldsymbol{\Phi}^{-1} \mathbf{1}_a)$ that has the unit length. Note that that $\mathbf{X}_a \mu = \alpha \mathbf{1}_a$ (angles between all inputs in A and μ are equal).

f) Compute the step size, $\gamma = \min_{j \in A^C}^{+} \left\{ \dfrac{C_{\max} - c_j}{\alpha - \theta_j}, \dfrac{C_{\max} + c_j}{\alpha + \theta_j} \right\}$

where min$^+$ indicates considering only positive minimum values over possible j.

g) Compute θ_j, which is defined as the inner product between all inputs and μ such as,

$$\theta_j = \mathbf{X}^T \mu$$

h) Update $\hat{\mathbf{Y}}_+ = \hat{\mathbf{Y}} + \gamma\mu$.

Repeat a–h until all inputs join the active set A, or $\sum_j |\beta_j|$ exceeds the given threshold.

vector in this direction as computed in Table 4.4(e). The amount of movement along μ_1, denoted as γ_1, is computed by equation in Table 4.4(f). y_1 denotes the OLS estimate of desired response y with input x_1. Note that the estimate by LAR ($\hat{y}_1$) moves toward y_1, but does not reach it. The next direction μ_2 intersects the angle between x_1 and x_2 (equiangular vector in the 2D space of x_1 and x_2) such that the angle between x_1 and the updated residual ($r_1 = y_2 - \gamma_1\mu_1$) is same as the one between x_2 and r_1. Because every input is standardized such that the correlation that is measured by the inner product of x_1 and x_2 can be estimated by the angle between x_1 and x_2, these two variables have the same absolute correlation with r_1 following the equation in Table 4.4(a). The coefficient γ_2 is computed again following Table 4.4(f), such that x_3 has the same absolute correlation with the next residual $r_2 = y_3 - (\gamma_1\mu_1 + \gamma_2\mu_2)$ as x_1 and x_2. So, the next direction μ_3 intersects the angle between x_1, x_2, and x_3. This procedure is repeated until the L_1 norm of coefficients reaches a given threshold.

LAR can be easily modified to implement LASSO; when some coefficients cross zero at a given step, those are forced to be zero. And the corresponding inputs are removed from the selected

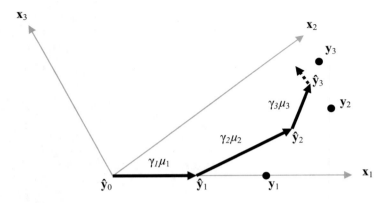

FIGURE 4.10: An illustration of the LAR procedure.

joint set. The LAR procedure can be continued with the remaining inputs because they still have the same absolute correlation with the current residual.

There are three major considerations in the implementation of LAR. First, LAR assumes the linearly independence between input variables. Second, the determination of threshold for the L_1 norm of coefficients is an open data-dependent problem. The performance of the linear model learned by LAR can be greatly influenced by a choice of this threshold. Finally, LAR was originally derived for static data. If we attempt to apply LAR to a linear model in BMIs, we have to cope with these three difficulties. Indeed, the embedded inputs are likely to be correlated with each other (although they might be linearly independent), so that LAR might not be able to operate optimally. Also, finding an optimal threshold will be a nontrivial task, but we may devise a surrogate data method to deal with this difficulty. The third issue is critical because the procedure has to be applied online for channel selection, and we will start by addressing our implementation [6] that can have wide applicability to time series modeling (not only BMIs).

Online Variable Selection. The LAR algorithm has been applied to variable selection in multivariate data analysis (static problems), but because LAR selects input variables by computation of correlation using the entire data set, it requires the assumption of stationary statistics when applied to time series. Therefore, a modified version of LAR, which selects a subset of input variables locally in time without the stationary assumption, has been proposed in Kim et al. [6]. Moreover, this algorithm provided can be computed online, and provides a real-time variable selection tool for the time-variant system identification problems.

Correlation between neuronal inputs and the desired response (hand trajectory) can be accomplished by recursively updating the correlation vector. The input covariance matrix can also be estimated recursively. By decoupling the variable selection part from the model update part in LAR, the input variables can be selected locally with recursive estimates of correlations. The modification to the LAR procedure using these recursions is described next.

The correlation in step a in Table 4.4 at a given step can be simply updated without computing residuals,

$$c_j(k) = c_j(k-1) - \gamma\theta_j \qquad (4.36)$$

for the kth step of variable selection. Hence, the update procedure of step h of Table 4.4 can be removed. Instead of computing the correlation with entire data, we can recursively estimate the correlation using a forgetting factor ρ, given by

$$p(n) = \rho p(n-1) + d(n)x(n) \qquad (4.37)$$

where $x(n)$ is an $1 \times M$ input vector at time instance n. $p(n)$ is utilized by the LAR routine such that $c_j(0) = p_j(n)$. For the computation of the covariance matrix Φ in step d in Table 4.4, we also estimate the input covariance matrix as

$$R(n) = \rho R(n-1) + x(n)^T x(n) \qquad (4.38)$$

This matrix is not directly used because Φ is the covariance of only a subset of inputs. Also, the input vectors are multiplied by the sign of correlations before computing Φ. Therefore, we need to introduce a diagonal matrix S whose elements are signs of $c_j(k)$ for $j \in A$. Then Φ can be computed using $R(n)$ and S as,

$$\Phi = SR_a S \qquad (4.39)$$

where R_a is an $L_a \times L_a$ (L_a is the length of A) matrix representing covariance among the selected input variables. R_a can be given by the elements of $R(n)$, that is, r_{ij} for $i, j \in A$. To remove the computation of the equiangular vector μ that requires a batch operation, we incorporate step e in Table 4.4 into step g in Table 4.4 such that

$$\theta_j = X^T \mu = X^T X_a (\alpha\Phi^{-1} \mathbf{1}_a) = \alpha X^T X_a \Phi^{-1} \mathbf{1}_a \qquad (4.40)$$

By noting that $X^T X_a$ is the jth columns of $R(n)$, for $j \in A$ followed by multiplication with S, we define R_{acol} to be a submatrix of $R(n)$ consisting of the jth columns for $j \in A$. Then step g in Table 4.4 can be computed as

$$\theta_j = \alpha R_{acol} S\Phi^{-1} \mathbf{1}_a \qquad (4.41)$$

TABLE 4.5: Online variable selection algorithm (taken from [30])

Given an $N \times M$ input matrix $\mathbf{X}$ (each row being M-dimensional sample vector), and an $N \times 1$ desired response matrix $\mathbf{Y}$, initialize $p(0) = 0$ and $\mathbf{R}(0) = 0$

Transform $\mathbf{X}$ and $\mathbf{Y}$ such that

$$\frac{1}{N}\sum_{i=1}^{N} x_{ij} = 0, \ \frac{1}{N}\sum_{i=1}^{N} x_{ij}^2 = 1, \text{ and } \frac{1}{N}\sum_{i=1}^{N} y_i = 0 \text{ for } j = 1, \ldots, M.$$

Update the correlation: $p(n) = (1 - \rho)p(n - 1) + \rho(d(n)x(n))$

Update the input covariance: $\mathbf{R}(n) = (1-\delta)\mathbf{R}(n - 1) + \delta(x(n)^{\mathrm{T}}x(n))$

Let $\mathbf{c}(0) = p(n)$.

For $k = 0, \ldots, M - 1$

Let $C_{\max} = \max_{j} \left\{ |c_j| \right\}$, and $A = [j: |c_j(\mathrm{k})| = C_{\max}]$.

Compute a diagonal matrix $\mathbf{S}$ with elements of sign of $c_j(k)$ for $j \in A$.

$$\Phi = \mathbf{S}\mathbf{R}_a\mathbf{S},$$

where $\mathbf{R}_a$ is submatrix of $\mathbf{R}(n)$ with jth rows and jth columns for $j \in A$.

$$\alpha = (\mathbf{1}_a^{\mathrm{T}}\Phi^{-1}\mathbf{1}_a)^{-1}$$

$$\theta_j = \alpha \, \mathbf{R}_{acol}\mathbf{S}\Phi^{-1}\mathbf{1}_a,$$

where $\mathbf{R}_{acol}$ is a matrix consisting of jth columns of $\mathbf{R}(n)$ for $j \in A$.

Compute the step size, $\gamma = \min_{j \in A^C}^{+} \left\{ \dfrac{C_{\max} - c_j}{\alpha - \theta_j}, \dfrac{C_{\max} + c_j}{\alpha + \theta_j} \right\}.$

Update correlation: $c_j(k) = c_j(k) - \gamma\theta_j$.

Hence, using Φ obtained by $\mathbf{R}(n)$ and $\mathbf{S}$, we can compute α and consecutively θ_j for $j \in A$. This modification removes the computation of the equiangular vector in step e in Table 4.4, which is not directly required for computing θ_j or γ.

In this way, $\mathbf{R}(n)$ and $\mathbf{p}(n)$ are estimated with the current input–output samples and fed into the modified LAR routine. Through this routine, a subset of input variables is selected with certain threshold for the L_1 norm of the coefficient vector. Therefore, it is possible to estimate which input variables are more correlated with the current desired response at every time instance. The procedure of this online variable selection is described in Table 4.5.

4.2.3 Real-Time Input Selection for Linear Time-variant MIMO Systems

In the linear MIMO system considered in this section, system outputs are assumed to be linearly and causally correlated with the neuronal spatiotemporal pattern of input channels. The temporal pattern of each input channel is represented by embedding input time series using a time delay line. In our representation, only a discrete time series is considered. Hence, an input sample is delayed by a discrete time delay operator, denoted by z^{-1} in the Z-domain. Let $x_j(n)$ be an input sample for the jth channel at time instance n. The temporal pattern of the jth channel is represented in the embedded vector space as $\boldsymbol{x}_j(n) \in \mathfrak{R}^L$, where L is the number of taps in the delay line. The linear approximation of the time-varying input–output mapping between all input channels and desired output signals, denoted by $y(n)$, is given by

$$y(n) = \sum_{j=1}^{M} \boldsymbol{w}_j(n)^T \boldsymbol{x}_j(n) + b(n) \qquad (4.42)$$

where $\boldsymbol{w}_j(n) \in \mathfrak{R}^L$ is the coefficient vector for the jth channel at time instance n, $b(n)$ is bias, and M is the number of channels. Note that the operation of $\boldsymbol{w}_j(n)^T \boldsymbol{x}_j(n)$ can be considered as filtering the jth channel input signal by a linear time-variant FIR filter. Therefore, $\sum_j \boldsymbol{w}_j(n)^T \boldsymbol{x}_j(n)$ can be the sum of the FIR filter outputs from every channel. We can also remove the bias term $b(n)$ by normalizing input and output, forcing them to have zero-mean.

The coefficient vector $\boldsymbol{w}_j(n)$ can be adapted by a number of methods, among which the LMS algorithm plays a central role because of its computational simplicity and tracking capability. With the LMS algorithm, the coefficient vector is updated as

$$\boldsymbol{w}_j(n+1) = \boldsymbol{w}_j(n) + \eta e(n)\boldsymbol{x}_j(n) \qquad (4.43)$$

for $j = 1, \ldots, M$. η is the learning rate, controlling convergence speed and the misadjustment, and $e(n)$ is the instantaneous error such that

$$e(n) = y(n) - \hat{y}(n) = y(n) - \sum_{j=1}^{M} \boldsymbol{w}_j(n)^T \boldsymbol{x}_j(n). \qquad (4.44)$$

See [31] for the review of the tracking linear time-variant systems using LMS.

Assuming linear independence between input channels in MIMO systems, we can apply on-line variable selection to channels instead of each tap output. To do so, we need to identify a variable that can represent the temporal patterns of input time series at each channel. If we consider learning a linear MIMO system, the estimation of desired signal is simply the sum of the FIR filters outputs from each channel. These individual filtered outputs indicate the relationship between desired outputs and the filtered channel input. Because online variable selection operates based on correlation, the filter output is hypothesized to be a sufficient variable to provide the correlation information between desired outputs and input temporal patterns

Hence, we choose the filter outputs as input to the online variable selection procedure. Then, the remaining question is how to learn each channel filter in real time, and here we choose to utilize LMS because of its simplicity and reasonable tracking performance in nonstationary environments [14].

Figure 4.11 depicts the overall architecture of the proposed online input channel selection approach. The embedded input vector $x_j(n)$ at the jth channel is filtered by an FIR filter with a coefficient vector of $w_j(n)$, yielding the filter output vector $z(n) = [z_1(n), \ldots, z_M(n)]^{\mathrm{T}}$. The autocovariance matrix $\mathbf{R}(n)$ of $z(n)$, and the cross-correlation vector $p(n)$ between $z(n)$ and desired output $y(n)$, are recursively estimated by (4.37) and (4.38), respectively. Then, the online variable selection algorithm receives $\mathbf{R}(n)$ and $p(n)$ to yield an LAR coefficient vector $g(n) = [g_1(n), \ldots, g_M(n)]^{\mathrm{T}}$. Note that some of elements in $g(n)$ can be equal to zero because of the L_1-norm constraint in the LAR procedure (Figure 4.12). Because the estimate of desired output is here the weighted sum of the channel filter outputs, an instantaneous error becomes

$$e(n) = y(n) - \hat{y}(n) = y(n) - g(n)^T z(n) \tag{4.45}$$

The update of $w_j(n)$ can then be accomplished by

$$w_j(n+1) = w_j(n) + \eta e(n) g_j(n) x_j(n) \tag{4.46}$$

Note the difference between this update rule and the one in (4.43) by additional weights on the filter outputs.

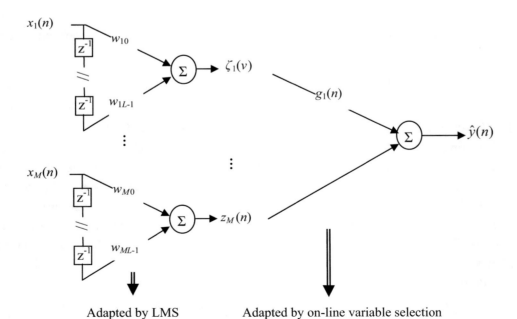

FIGURE 4.11: A diagram of the architecture of online channel selection method.

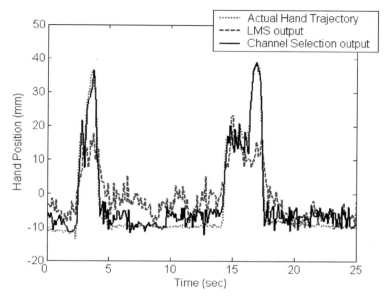

FIGURE 4.12: An example of the outputs of two tracking systems, SYS1 (dashed line) and SYS2 (solid line), on top of actual hand trajectory (dotted line).

Selection of the Threshold. The constraint on the L_1-norm of the LAR coefficients impacts the neuronal subset selection as follows: a smaller threshold of the constraint will result in a large subset, increasing a chance to select irrelevant channels, and a larger threshold will result in a smaller subset, increasing a chance to miss relevant channels. Hence, a careful approach to determine this threshold is critical. Towards this end, we propose using surrogate data to select the criterion parameters.

The surrogate data are generated to uncouple the hand movements from the neural activity using two different procedures: 1) shifting either the kinematics or the neural activity, or 2) shuffling the phase of the kinematic signals while preserving its power spectral density (PSD) to ensure the same second-order statistics after perturbation [32]. The hypothesis is that there is no correlation between the neural data and the hand movements in the surrogates. Therefore, the threshold should be set such that no subsets are found in the surrogate data sets.

The online variable selection includes two components: a threshold for the correlation between the model output $\hat{d}_{\text{LMS}}(n)$ and desired signal $d(n)$; and a threshold for the maximum correlation in the variable selection algorithm, that is $C_{\text{max}}(j)$. The first threshold, δ_1, plays a role such that the variable selection algorithm is not activated when a correlation between $\hat{d}_{\text{LMS}}(n)$ and $d(n)$ is lower than δ_1. In this case, no subset is selected at time n. This happens when the FIR filters are not sufficiently adapted to track the target system. Accordingly, unreliable subsets may result if the selection is performed on such inadequate filter outputs. On the other hand, when the correlation exceeds δ_1, the variable selection algorithm runs until $C_{max}(j)$ becomes lower than the second threshold, δ_2. We empirically search certain values of $[\delta_1, \delta_2]$ with which the probability of selecting at least one channel is very low for both surrogate data sets, but reasonably high for the original data.

To determine δ_1, a correlation between $\hat{d}_{LMS}(n)$ and $d(n)$ is recursively estimated such as

$$\xi(n) = \mu\xi(n-1) + \frac{\hat{d}_{LMS}(n)d(n)}{\sqrt{p(n)q(n)}} \quad , \tag{4.47}$$

where μ is a forgetting factor that is usually defined for the recursive least squares (RLS) [14]. $p(n)$ and $q(n)$ represent the power estimates for $\hat{d}_{LMS}(n)$ and $d(n)$, respectively. Normalization by the square root of $p(n)q(n)$ prevents $\xi(n)$ from being biased to a large magnitude of $d(n)$. The power is estimated through similar recursions:

$$p(n) = \mu p(n-1) + \hat{d}^2_{LMS}(n).$$
$$q(n) = \mu q(n-1) + d^2(n) \tag{4.48}$$

If $\xi(n) \geq \delta_1$, the online variable selection algorithm is activated. If $\xi(n) < \delta_1$, an empty subset is yielded and $\hat{d}_{LAR}(n)$ is set equal to $\hat{d}_{LMS}(n)$.

Once the online variable selection algorithm is started, δ_2 is used to stop the algorithm until $C_{max}(j) < \delta_2$ at the jth iteration. Here, we describe how $C_{max}(j)$ represents the correlation of inputs with a desired output. In the LAR, if two successively selected inputs, x_{j-1} and x_j, have similar correlations with a desired output, d, the decrease from $C_{max}(j-1)$ to $C_{max}(j)$ will be small. On the other hand, if x_{j-1} has more correlation than x_j, $C_{max}(j)$ will be much smaller than $C_{max}(j-1)$. This is illustrated in Figure 4.13. Consider the data $[\mathbf{X}, \mathbf{d}]$, where $\mathbf{X}$ is an input matrix whose rows are input samples, and $\mathbf{d}$ is an output vector. Suppose that $\mathbf{X}$ has two columns, x_1 and x_2. We assume that x_1, x_2, and $\mathbf{d}$ are standardized with zero mean and unit variance. Suppose x_1 has more correlation with $\mathbf{d}$. The LAR starts to move in the direction of x_1. It finds the coefficient β_1 for x_1 such that $|x_1^T r_1|$ = $|x_2^T r_1|$, where $r_1 = \mathbf{d} - \beta_1 x_1 \equiv \mathbf{d} - y_1$. Then, the maximum correlation changes from $C_{max}(0) = |x_1^T \mathbf{d}|$ to $C_{max}(1) = |x_1^T r_1| = |x_2^T r_1|$. From this, we can see that the angle between x_1 and r_1 is equal to the angle between x_2 and r_1—the equiangular property of the LAR. $C_{max}(j)$ is related with the angles such that $C_{max}(j) \approx \cos\theta_j$, where θ_j represents the angle at the jth step. The diagram on the left side of Figure 4.13 illustrates the case when x_1 and x_2 have similar correlations with $\mathbf{d}$. In this case, a small difference between θ_0 and θ_1 leads to a small decrease from $C_{max}(0)$ to $C_{max}(1)$. On the other hand, the diagram on the right side illustrates the case when x_2 is considerably less correlated with $\mathbf{d}$ than x. In this case, a large difference between θ_0 and θ_1 causes $C_{max}(1)$ to decrease significantly.

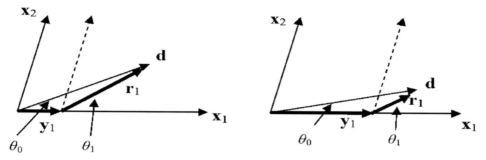

FIGURE 4.13: Illustration of the maximum correlation in the online variable selection algorithm. Note that **d** represents a desired output projected in the input space.

Therefore, the curve of $C_{max}(j)$ over iterations can represent the correlation between each selected input and a desired output.

4.3   EXPERIMENTAL RESULTS

We first determined two thresholds, δ_1 and δ_2, by applying the linear adaptive systems to both the original and surrogate data sets. The first surrogate data are composed of neuronal inputs and the shifted HPs. The HP data are shifted by 5 sec for a hand reaching task because successive reaching movements has an interval of approximately 10 sec. For the 2D cursor control task, the HP data are shifted by 10 sec to sufficiently destroy synchronization. The second surrogate data consist of neuronal inputs and the shuffled hand trajectory data in which the phase of the HP signals are shuffled while preserving PSD.

Figure 4.14 shows an example of the curve of $C_{max}(j)$ for the case when we use the original data and the surrogate data, respectively. In this example, $C_{max}(j)$ for both the original and the surrogate data were evaluated at the same time instance. A large difference between the C_{max} curves can be observed; for example, three channels were selected for the original data whereas only one channel was selected for the surrogate data. The dotted line indicates the threshold of selection.

The linear adaptive systems were implemented for 2000-sec (20 000 samples three-dimensional (3D) data and 1600-sec (16 000 samples) 2D data sets. The online variable selection algorithm starts after 100 sec for the 3D data, and 400 sec for the 2D data, in order to allow the LMS to sufficiently adapt filter weights in the beginning. A learning rate of LMS is set such that the sum of the FIR filter outputs without subset selection can track the hand trajectory reasonably well in the original data. Note that it must not be set too large, otherwise the filter weights change too rapidly over time, which will lead to a large misadjustment; therefore, the estimation of the correlation of individual neuronal channels from the filter outputs becomes unreliable.

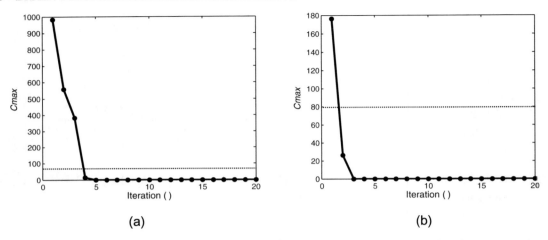

FIGURE 4.14: Examples of the maximum absolute correlation curves over variable selection iterations for (a) the original data and (b) the surrogate data.

From experiments with the surrogate data sets, the first threshold, δ_1, is determined such that few neuronal channels are selected in the surrogate data. The model parameters and the thresholds are summarized in Table 4.6. Note that δ_1 is determined much lower for more complex target hitting data, indicating that the correlation of the FIR filter outputs with kinematics tends to be lower for these data.

We show examples of neuronal subset selection results for the original and surrogate hand reaching data, with the thresholds setting as described earlier, in Figure 4.15. Figure 4.15a shows

TABLE 4.6: The model parameters and neuronal subset selection thresholds		
PARAMETER	**3D**	**2D**
A learning rate for NLMS; η in (4.46)	0.1	0.3
A forgetting factor for FIR filter correlation; μ in (4.47)	0.95	0.95
A forgetting factor for channel covariance; ρ in (4.38)	0.8	0.9
A threshold for activation of variable selection; δ_1	0.7	0.3
A threshold for correlation in variable selection; δ_2	80	80

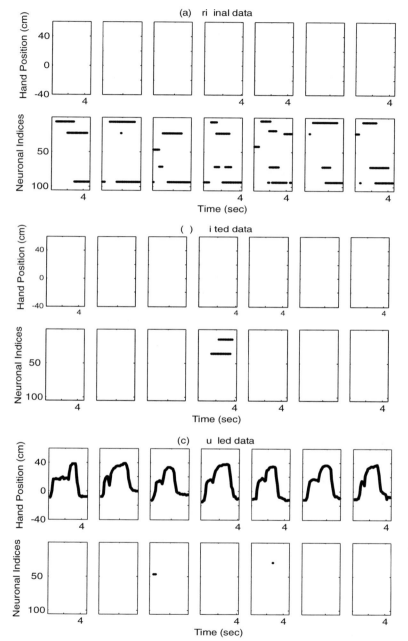

FIGURE 4.15: Neuronal subset examples for the identical 3D reaching movements using different data: (a) the original data; (b) the shifted data by 5 sec; (c) the shuffled data. Black dots in the below figures denote the selected neuronal channels.

the examples of neuronal subset selection for the original data. Figure 4.15b and c show the examples for the shifted data and the shuffled data, respectively. The neuronal subsets are presented along with the corresponding HPs (z coordinate) of seven reaching movement examples. It is clear that very few subsets are selected for the surrogate data. To quantify these subset selection results, we define a selection rate as the average number of neuronal channels per time instance. From the experimental results, the selection rates were 0.006 ± 0.006 for the original hand reaching data, 0.001 ± 0.003 for the shifted data, and $9.1 \times 10^{-4} \pm 0.004$ for the shuffled data, respectively. For 2D cursor control data, the selection rates were 0.015 ± 0.009 for the original data, and 0.002 ± 0.006 for the shifted data (the selection rate for the shuffled data is considerably lower). Therefore, these results demonstrate that we can determine the thresholds using the surrogate data with which neuronal subsets selected from the synchronized data is determined to represent real-time correlations of neuronal activities and kinematics.

We examined if the neuronal subsets showed dependency on the initial condition of the linear adaptive system. We first define a selection vector as

$$s(n) = [s_1(n), s_2(n), \cdots, s_M(n)]^T$$

(4.49)

where $s_j(n) = 1$ if the jth neuronal channel was selected at the nth bin, and $s_j(n) = 0$ otherwise. We repeated the neuronal subset selection analysis 100 times for different initial filter weights. Then, an average of $s(n)$, denoted as $\underline{s}(n)$, over 100 repetitions was computed. If the linear adaptive system is robust to initial conditions, the nonzero elements of $\underline{s}(n)$ should be close to 1. Hence, we examined the statistics of those nonzero elements. For 3D food reaching data, the average and standard deviation were 0.85 and 0.26, respectively. However, if we are only concerned with the subset selection for movements, the average and standard deviation estimated only during movements were 0.94 and 0.15, respectively. Because we are typically more interested in neuronal subsets during given movements, the latter result demonstrates that the linear system is reasonably robust to initial conditions. For 2D cursor control data, the average and standard deviation of the nonzero elements of $\underline{s}(n)$ were 0.91 and 0.19, respectively, which again demonstrates the robustness to initial conditions.

To find neuronal subsets without separation into each spatial coordinate, we performed the analysis for individual coordinates and combined the resulting subsets into one. To that end, we performed a Boolean OR operation with $s(n)$ resulted from every coordinate analysis. This operation yields a single joint neuronal subset at each time instance. In other words, if a channel is selected for at least one coordinate of HP, it belongs to the joint subset.

To investigate the temporal variation of neuronal subsets, we tracked the joint subsets over the entire data. To focus the analysis only on the movement portion of the 3D data, we divided the entire neuronal subsets into individual segments corresponding to each reaching movement. This resulted in 149 movement segments. In each segment, we collected neuronal channels that were selected at least for two consecutive bins. Figure 4.16 displays this analytical result. From this figure, we can observe which neurons were consistently selected through the entire set of reaching movements. For instance, neurons indexed by 5, 7, 23, 29, and 93 are shown to be selected in most segments. On the other hand, neurons indexed by 19, 45, 70, 71, and 84 are partially selected in some segments. In particular, neuron 71 seems to be consistently selected in the late segments.

This observation is further demonstrated in Figure 4.17. In this figure, we displayed neuronal subset examples in the early movement segments (485, 509, and 600 sec from the beginning of subset selection) and the late movement segments (1666, 1713, and 1781 sec). It can be observed from this figure that there are neurons that are consistently selected such as 5, 7, and 93 for both early and late segments. However, some neurons selected in the early segments are not selected in the late segments. For instance, neuron 70 is no longer selected in the late segments. Also, other neurons (such as 71) are not selected in the early segments but selected in the late segments. It is interesting to see this transition of a correlation with kinematics from the neuron 70 to 71 through time, because the activities of those neurons are collected from the adjacent electrodes in the PMd area.

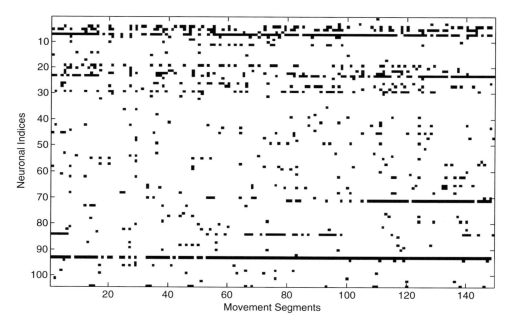

FIGURE 4.16: Temporal patterns of neurons selected during 3D food reaching movements. Black dots represent selection of neurons for each movement segment (see text). The segment indices are sequentially ordered in time.

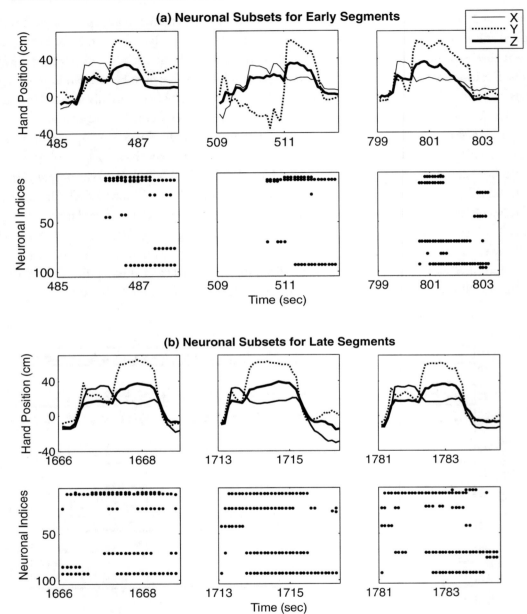

FIGURE 4.17: Neuronal subset examples for the entire 3D food reaching movement segments.

Similar observations were made for the 2D cursor control data. In Figure 4.18, neuronal subset examples for five segments with specific hand movement are displayed. In these segments, the animal moved the hand in the positive diagonal direction; that is, the hand was moved back and forth in 270° and 45° for about 2.5–5 sec. From these examples, we can observe neurons that are consistently selected such as 69, 80, 84, 92, 99, 108, and 110. However, there are a number of neurons that are selected only in particular segments such as 45, 54, 67, 149, etc. Most of selected neurons were collected from the M1 area, yet some neurons from the PMd area join the subsets in the last two segments (late parts of the data set).

When we compared neuronal subsets from our analysis to the sensitivity analysis obtained from the trained models in the same data set [7], we were able to find consistency among neuronal contributions. For instance, neurons 5, 7, 23, 71, and 93 in the 3D food reaching data were observed in the top-ranked group of neurons sorted by sensitivity analysis. Also, neurons 54, 69, 80, 84, 92,

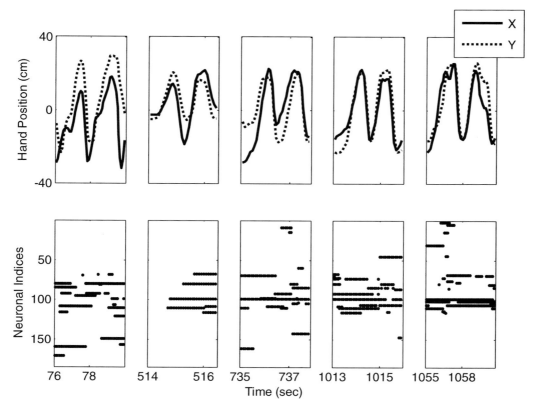

FIGURE 4.18: Neuronal subset examples for the 2D cursor control task.

99, 108, 110, 149, and 167 in the 2D target hitting data were similarly observed in the top-ranked sensitivity group. These comparisons show that neuronal subsets selected in our analysis significantly contribute to building decoding models (nonlinear as well as linear models). The advantage of this method would be the capability of detecting time-varying changes of composition of subsets.

To demonstrate the effect of the time-variant relationship between the neural activity and kinematics to the decoding, we performed a brief statistical test on the decoding performance using the linear filter. We first divided the entire data into three disjoint sequential segments, say, SEG1, SEG2, and SEG3 (such that SEG1 preceded SEG2 and SEG3). Then, the first linear filter was built using samples randomly drawn from SEG1, where the Wiener–Hopf solution was used to learn the filter coefficients [14]. The second linear filter was built using samples randomly drawn from SEG2. After building these filters, we drew samples randomly from the last segment SEG3, and used them to test each filter's decoding accuracy. The accuracy was measured by the mean absolute error (MAE), which indicates the mean distance between the actual hand trajectory and the estimated trajectory. We hypothesized that the decoding accuracy of two filters should be similar to each other if the relationship of neural activity to kinematics is unchanged. We repeated this building and decoding procedure multiple times, and obtained a set of MAE values for each filter. Then, a nonparametric statistical method (the Kolmogorov–Smirnov (KS) test) was applied to test if two empirical distributions of the MAE from each filter were statistically equal. Specifically, the null hypothesis of the KS test was given by,

$$H_0 : F_1(|e|) \geq F_2(|e|),$$
(4.50)

where $F_i(|e|)$ is an empirical cumulative density function (CDF) of the MAE from SEG1 ($i = 1$) or SEG2 ($i = 2$). If H_0 is rejected, the alternative hypothesis is then $F_1(|e|) < F_2(|e|)$, which indicates that the second linear filter yielded less errors.

The size of SEG1 or SEG2 was 800 sec long for the 3D data and 700 sec for the 2D data. We randomly drew 4000 samples from each segment to build the linear filter. The size of SEG3 was 400 sec long for the 3D data and 200 sec for the 2D data. We randomly drew 500 samples from SEG3 to test the linear filter. The procedure was repeated 1000 times. The empirical statistics are described in Table 4.7, in which it is clearly shown that the second linear filter decoded kinematics

TABLE 4.7: Average MAE using two linear filters (in centimeters)		
	3D	2D
First filter (SEG1)	32.83 ± 3.43	19.39 ± 0.47
Second filter (SEG2)	17.94 ± 0.61	18.64 ± 0.37

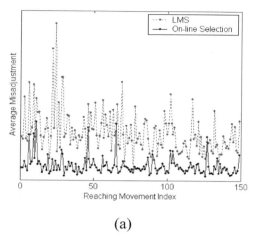
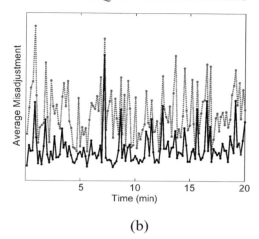

(a)   (b)

FIGURE 4.19: A comparison of misadjustments between the online neuronal channel selection and the NLMS: (a) 3D data, (b) 2D data.

from neural activity for samples in SEG2 better than the first filter. The statistical test showed that the null hypothesis was rejected for both 3D and 2D data sets ($P < 10^{-12}$). These results imply that we may need to rebuild the linear filter during the session to track the change of the relationship between neural activity and kinematics.

Finally, we investigated the misadjustment of two linear adaptive systems, one of which was adapted using the NLMS and the online subset selection algorithm, whereas the other was adapted using only the NLMS. For the 3D data, we computed the average misadjustment for each movement segment. For the 2D data in which the continuous movement was recorded, we arbitrarily divided the data into 10-sec segments and computed the average misadjustment for each segment. Figure 4.19 shows the results of these average misadjustments for each data set. In this figure, the linear adaptive system with the online subset selection exhibits smaller misadjustments than the normal linear adaptive system for almost every segment. This result is consistent with a previous demonstration of the superior tracking performance of the online subset selection algorithm, as reported in [33].

4.4 SUMMARY

A reduction in the number of free parameters without affecting performance leads directly to BMI systems that require less power, bandwidth, and computational demands. Solving these challenges will bring us one step closer to real, portable BMIs. Regularization and analysis of the most important neurons in the model-dependent methods has opened up directions to better understand how neural activity can be effectively used for BMI design. Since the work of Wessberg et al, it is known that the performance of linear models for BMIs improves as more neurons are recorded, but that the performance improvement must be coupled with the *right* neurons. Irrelevant neurons can add to the model bias and variance, thus reducing performance. One must also remember that if channel selection is used the reduced number of cells makes BMIs more sensitive to the instability of neuronal firings over time. Recent studies are showing that the activity of individual neurons and cortical areas used in BMI experiments can vary considerably from day to day [34]; therefore, the variance over time of the importance of neurons must be quantified in future studies. It is therefore not clear how practical the neuron selection techniques will be in the surgery stage. For these reasons, we advocate the use of a higher sampling of the cortical activity to help improve this ratio until other models are proposed that take advantage of the information potentially contained in the spike trains and not exploited by linear models.

In this model-based framework, it is clear that better BMIs can be built by combining regularization techniques with a subset of important cells. This question is rooted in our goal to build models for BMIs that generalize optimally. It should not be forgotten that regularization and channel selection is biased by the type of model chosen, its performance level, and by noise in the data. Therefore, it is important to pursue a model-independent approach to preprocess the data. We hypothesized from a combined neurophysiological and modeling point of view that highly modulated neurons spanning the space of the kinematic parameter of interest should be chosen. Intuitively, these constraints make sense for the following reasons: If a cell is modulated to the kinematic parameter, the adaptive filter will be able to correlate this activity with the behavior through training. Otherwise, neural firing works as a broadband excitation that is not necessarily related to better performance. If a group of cells are highly modulated to only a part of the space, the adaptive filter may not be able to reconstruct data points in other parts of the space.

A final comment goes to the overall performance of the BMI system built from adaptive models. Although it is impressive that an optimal linear or nonlinear system is able to identify the complex relations bnetween spike trains in the motor cortex and hand movements/gripping force, a correlation coefficient of ~0.8 may not be sufficient for real-world applications of BMIs. Therefore, further research is necessary to understand what is limiting the performance of this class of adaptive linear and nonlinear systems. Another issue relates to the unrealistic assumption of stationarity in neural firings over recording sessions that is used to derive results presented in this. In future studies, it will be necessary to assess the time variability of the neuronal rankings and determine its effect on model generalization.

REFERENCES

1. GemanS., E. Bienenstock, E., and R. Doursat, *Neural networks and the bias/variance dilemma.* Neural Computation, 1992. **4**: pp. 1–58.
2. Sanchez, J.C., et al., *Input–output mapping performance of linear and nonlinear models for estimating hand trajectories from cortical neuronal firing patterns*, in International Work on Neural Networks for Signal Processing. 2002. IEEE, Martigny, Switzerland. doi:10.1109/NNSP.2002.1030025

3. Sanchez, J.C., et al., *A comparison between nonlinear mappings and linear state estimation to model the relation from motor cortical neuronal firing to hand movements*, in SAB Workshop on Motor Control in Humans and Robots: on the Interplay of Real Brains and Artificial Devices. 2002. University of Edinburgh, Scotland.

4. Sanchez, J.C., et al., *Interpreting neural activity through linear and nonlinear models for brain machine interfaces*, in International Conference of Engineering in Medicine and Biology Society. 2003. Cancun, Mexico. doi:10.1109/IEMBS.2003.1280168

5. Wahba, G., Spline Models for Observational Data. 1990, Montpelier: Capital City Press.

6. Kim, S.-P., J.C. Sanchez, and J.C. Principe, *Real time input selection for linear time-variant MIMO systems.* Optimization Methods and Software, 2007. **22**: pp. 83–98. doi:10.1080/105 56780600881886

7. Sanchez, J.C., et al., *Ascertaining the importance of neurons to develop better brain machine interfaces.* IEEE Transactions on Biomedical Engineering, 2003. **61**(6): pp. 943–953. doi:10.1109/TBME.2004.827061

8. Hadamard, J.P.U.B., *Sur les problèmes aux dérivées partielles et leur signification physique.* Princeton University Bulletin, 1902: pp. 49–52.

9. Tikhonov, A. and V. Arsenin, Solution of Ill-Posed Problems. 1977, Washington: Winston.

10. Neal, R., Bayesian Learning for Neural Networks. 1996, Cambridge: Cambridge University Press.

11. Vapnik, V., The Nature of Statistical Learning Theory. Statistics for Engineering and Information Science. 1999, New York: Springer-Verlag.

12. Stewart, G.W., Introduction to Matrix Computations. 1973, New York: Academic Press.

13. Klema, V.C. and A.J. Laub, *The singular value decomposition: Its computation and some applications.* IEEE Transactions on Automatic Control, 1980. **AC-25**: pp. 164–176. doi:10.1109/TAC.1980.1102314

14. Haykin, S., Adaptive Filter Theory. 3rd ed. 1996, Upper Saddle River, NJ: Prentice-Hall International.

15. Hoerl, A.E. and R.W. Kennard, *Ridge regression: Biased estimation for nonorthogonal problems.* Technometrics, 1970. **12**(3): pp. 55–67.

16. Weigend, A.S., D.E. Rumelhart, and B.A. Huberman, *Generalization by weight–elimination with application to forecasting.* Advances in Neural Information Processing Systems 3. R.P. Lippmann, J. Moody, and D.S. Touretzky, eds. 1991. pp. 875–882, Morgan Kaufmann: San Mateo, CA.

17. Larsen, J., et al., *Adaptive regularization in neural network modeling*, Neural Networks: Tricks of the Trade, G.B. Orr and K. Muller, eds., 1996, Germany: Springer, pp. 113–132. doi:10.1007/3-540-49430-8_6

18. Geisser, S., *The predictive sample reuse method with applications.* Journal of the American Statistical Association, 1975. **50**: pp. 320–328. doi:10.2307/2285815

19. Principe, J.C., B. De Vries, and P.G. Oliveira, *The gamma filter—a new class of adaptive IIR filters with restricted feedback.* IEEE Transactions on Signal Processing, 1993. **41**(2): pp. 649–656. doi:10.1109/78.193206

20. Sandberg, I.W. and L. Xu, *Uniform approximation of multidimensional myopic maps.* IEEE Transactions on Circuits and Systems, 1997. **44**: pp. 477–485.

21. De Vries, B. and J.C. Príncipe, *The gamma model: a new neural network model for temporal processing.* Neural Networks, 1993. **5**: pp. 565–576.

22. de Jong, S., *SIMPLS: An alternative approach to partial least squares regression.* Chemometrics and Intelligent Laboratory Systems, 1993. **18**: pp. 251–263. doi:10.1016/0169-7439(93)85002-X

23. Stone, M. and R.J. Brooks, *Continuum regression: cross-validated sequentially constructed prediction embracing ordinary least squares, partial least squares and principal components regression (with discussion).* Journal of Royal Statistical Society, Series B, 1990. **52**: pp. 237–269.

24. Kim, S.P., et al. *A hybrid subspace projection method for system identification*, in Proceedings of the International Conference on Acoustics, Speech, and Signal Processing. 2003. doi:10.1109/ICASSP.2003.1201683

25. Fetz, E.E., *Are movement parameters recognizably coded in the activity of single neurons.* Behavioral and Brain Sciences, 1992. **15**(4): pp. 679–690.

26. Fu, L. and T. Chen. *Sensitivity analysis for input vector in multilayer feedforward neural networks*, in IEEE International Conference on Neural Networks. 1993. San Francisco, CA. doi:10.1109/ICNN.1993.298559

27. Hastie, T., R. Tibshirani, and J. Friedman, Elements of statistical learning: data mining, inference and prediction. 2001, New York: Springer-Verlag.

28. Tibshirani, R.J., *Regression shrinkage and selection via the lasso.* Royal Statististical Society B, 1996. **58**(1): pp. 267–288.

29. Efron, B., et al., *Least angle regression.* Annals of Statistics, 2004. **32**: pp. 407–499.

30. Kim, S.-P., et al., *Tracking multivariate time-variant systems based on on-line variable selection*, in IEEE Workshop on Machine Learning for Signal Processing. 2004: Sao Luis, Brazil.

31. Haykin, S., et al., *Tracking of Linear Time-Variant Systems*, in IEEE MILCOM. 1995.

32. Palus, M. and D. Hoyer, *Detecting nonlinearity and phase synchronization with surrogate data.* IEEE Engineering in Medicine Biology Magazine, 1998. **17**(6): pp. 40–45. doi:10.1109/51.731319

33. Kim, S.P., J.C. Sanchez, and J.C. Principe, *Real time input subset selection for linear time-variant MIMO systems.* Optimization Methods and Software, 2007. **22**(1): pp. 83–98.

34. Carmena, J.M., et al., Learning to control a brain–machine interface for reaching and grasping by primates. PLoS Biology, 2003. **1**: pp. 1–16.

• • • •

CHAPTER 5

Neural Decoding Using Generative BMI Models

Additional Contributors: Yiwen Wang and Shalom Darmanjian

This chapter will address generative models for BMIs, which are a more realistic modeling approach because they take into consideration some of the known features of motor cortex neurons. This chapter still uses neuronal firing rates as inputs to the model, but the mathematical foundations can also be applied, with appropriate adaptations, to spike trains as will be discussed in the next chapter. Generative models are more in tune with the physical neural systems that produce the data and therefore are examples of "gray box" models. Three examples of "gray box" models can be found in the BMI literature. One of the most common examples is Georgopoulos population vector algorithm (PVA) [1] as we briefly mentioned in Chapter 3. Using observations that cortical neuronal firing rates were dependent on the direction of arm movement, a model was formulated to incorporate the weighted sum of the neuronal firing rates. The weights of the model are then determined from the neural and behavioral recordings. A second example is given by Todorov's work who extended the PVA by observed multiple correlations of M1 firing with movement position, velocity, acceleration, force exerted on an object, visual target position, movement preparation, and joint configuration [2–13]. With these observations, Todorov proposed a minimal, linear model that relates the delayed firings in M1 to the sum of many mechanistic variables (position, velocity, acceleration, and force of the hand) [14]. Todorov's model is intrinsically a *generative model* [15, 16]. Using knowledge about the relationship between arm kinematics and neural activity, the *states* (preferably the feature space of Todorov) of linear or nonlinear *dynamical* systems can be assigned. This methodology is supported by a well known training procedure developed by Kalman for the linear case [17], and has been recently extended to the nonlinear case under particle filters [18] and other graphical models or Bayesian network frameworks [19]. Because the formulation of generative models is recursive in nature, the model is well suited for learning about motor systems because the states are all intrinsically related in time. The third modeling approach is the hidden Markov model (HMM), a graphical model where the dependencies over time are contained in the present state, and the observations are assumed to be discrete. HMM models are the leading technology in speech recognition because they are able to capture very well the piecewise nonstationary of speech [20]. It turns out that speech production is ultimately a motor function, so HMMs can potentially be also very useful for motor BMIs. One of the major differences is that there is a well established finite repertoire of speech atoms called the phonemes, which allow for the use of the HMM framework. Research into animal motor control is less advanced, but recently there has been interest in studying complex motor actions as a succession of simpler movements that can be properly called "movemes" [21]. If this research direction meets the expectation, then HMMs would play a more central role in motor BMIs because the model building problem would be similar to speech recognition.

5.1   POPULATION VECTOR CODING

As presented in the Introduction, the population vector algorithm is a physiologically based model that assumes that a cell's firing rate is a function of the velocity vector associated with the movement performed by the individual. Population vector coding is based on the use of tuning curves [22], which in principle provide a statistical relationship between neural activity and behavior. The tuning or preferred direction [2] of each cell in the ensemble convey the expected value of a probability density function (PDF) indicating the average firing a cell will exhibit given a particular movement direction. The PVA model relating the tuning to kinematics is given by (5.1)

$$s_n(\mathbf{V}) = b_0^n + b_x^n v_x + b_y^n v_y + b_z^n v_z = \mathbf{B} \cdot \mathbf{V} = |\mathbf{B}||\mathbf{V}| \cos\theta \, , \qquad (5.1)$$

where the firing rate s for neuron n is a weighted $(b_{x,y,z}^n)$ sum of the components $(v_{x,y,z})$ of the unit velocity vector $\mathbf{V}$ of the hand plus mean firing rate b_n^0. The relationship in (5.1) is the inner product between the velocity vector of the movement and the weight vector for each neuron, that is, the population vector model considers each neuron independently of the others (see Figure 5.1). The inner product (i.e., spiking rate) of this relationship becomes maximum when the weight vector $\mathbf{B}$ is collinear with the velocity vector $\mathbf{V}$. At this point, the weight vector $\mathbf{B}$ can be thought as the neuron's preferred direction for firing because it indicates the direction for which the neuron's activity will be maximum. The weights b_n can be determined by least squares techniques [1]. Each neuron makes a vector contribution w in the direction of P_i with magnitude given in (5.2), where b_0 is the mean firing rate for neuron n. The resulting population vector or movement is given by (5.3), where the reconstructed movement at time t is simply the sum of each neuron's preferred direction weighted by the firing rate.

$$w_n(\mathbf{V}, t) = s_n(\mathbf{V}) - b_0^n \, , \qquad (5.2)$$

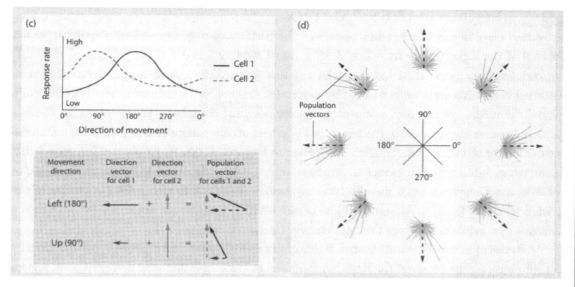

FIGURE 5.1: Illustration of the mechanisms of population vector approach.

$$\mathbf{P}(\mathbf{V},t) = \sum_{n=1}^{N} w_n(\mathbf{V},t)\frac{\mathbf{B}_n}{\|\mathbf{B}_n\|}, \qquad (5.3)$$

The population vector can therefore be thought of as a local linear model, where the contributions of each neuron are weighted appropriately from the trajectory. It should be noted here that the PVA approach includes several assumptions whose appropriateness in the context of neural physiology and motor control will only be briefly considered here (see Ref. [23] for a full discussion). First, each cell is considered independently in their contribution to the kinematic trajectory. The formulation does not consider feedback of the neuronal firing patterns; a feature found in real interconnected neural architectures. Second, neuronal firing counts are linearly combined to reproduce the trajectory. At this moment it is unclear how the neural activation of nonlinear functions can create arbitrarily complex movement trajectories. The PV model determines the movement direction from neural activity, but the reconstruction of hand trajectory also requires the estimation of the speed. Georgopoulos et al [1] directly used the magnitude of the population vector, $|\mathbf{P}(n)|$, for estimating the instantaneous speed, and Moran and Schwartz [24] extended the PVA model to include the hand speed. Then, the trajectory of the hand position was reconstructed by a time series of population vectors that were connected tip to tail as time progresses.

5.2   SEQUENTIAL ESTIMATION

A second class of models that has been used for BMIs uses a generative, state space approach instead of the input-output modeling described in Chapters 3 and 4 [5, 16]. Generative models utilize a Bayesian formulation and they offer a general framework to estimate goal directed behavior encoded in multichannel spike trains. The movement kinematics are defined as state variables of a neural dynamical system from a sequence of noisy neuronal modulation observations. In this approach, the data are analyzed in the neural space instead of using an input-output model to map the neural inputs and desired kinematics. The generative approach creates an observation model that incorporates information from measurements using a recursive algorithm to construct the posterior PDF of the kinematics given the observations at each time. By estimating the expectation of the posterior density (or by maximum likelihood estimation), the movement kinematic estimate can be recovered probabilistically from the multichannel neural recordings.

Bayesian learning assumes that all forms of uncertainty should be expressed and measured by probabilities. At the center of the Bayesian formulation is a simple and very rich expression known as Bayes' rule. Given a data set $D = [u_{1:N}, z_{1:N}]$, and a set of possible descriptive data models M_k, $k = 0, \ldots, K$, the Bayes' rule becomes

$$p(M_i \mid D) = \frac{p(D \mid M_i)}{\sum_k p(D \mid M_k) p(M_k)} p(M_i) \qquad (5.4)$$

This rule simply states that the posterior distribution ($p(M|D)$) can be computed as the ratio of the likelihood ($p(D|M)$) over the evidence ($p(D) = \sum_k p(D \mid M_k) p(M_k)$) times the prior probability ($p(M)$). The resulting posterior distribution incorporates both a priori knowledge and the information conveyed by the data.

This approach can be very useful in parametric model building ($f(u,w)$), where u is the input and w are system parameters because one can assign prior probabilities not only to the parameters of the system but also to model order k as $p(w,k)$ and even to the noise. Once the data are passed thru the system we can build the posterior distribution $p(w,k \mid u_{1:N}, z_{1:N})$ using Bayes' rule, where z are the system outputs. Because the posterior embodies all the statistical information about the parameters and their number given the measurements and the prior, one can theoretically obtain all features of interest (marginal densities, model selection, and parameter estimation) by standard probability marginalization and transformation techniques. For instance, we can predict the value of the next system output by (5.5)

$$E(z_{N+1} \mid u_{1:N+1}, z_{1:N}) = \int \hat{f}(k, w, u_{N+1}) p(k, w \mid u_{1:N}, z_{1:N}) dk dw \qquad (5.5)$$

Note that the predictions must be based on *all* possible values of the network parameters weighted by their probability in view of the training data. Moreover, note that Bayesian learning is fundamentally based on *integration* over the unknown quantities. In (5.5), because we want to find the expected value of the output, which is a function of the model parameters and model order, we will have to integrate over these unknown quantities using all the available data. Because these integrals are normally multidimensional, this poses a serious problem that was only recently mitigated through the development of very efficient *sampling-based evaluation of integrals* called Monte Carlo sampling [25]. From a Bayesian perspective, Monte Carlo methods allow one to compute the full posterior probability distribution.

In BMIs we are interested in problems where the noisy neuronal recordings can be thought as the output of a system controlled by its state vector x, which could include kinematics such as position, velocity, acceleration, and force. Moreover, we assume that the system state vector may change over time as shown in Figure 5.2.

The state space representation of a discrete stochastic dynamical system is given by

$$\begin{cases} x_{t+1} = x_t + v_t \\ z_t = \hat{f}_t(u_t, x_t) + n_t \end{cases} \qquad (5.6)$$

where the noise term v_t is called the process noise and n_t the measurement noise. The first equation defines a first-order Markov process $p(x_{t+1}|x_t)$ (also called the system model), whereas the second defines the likelihood of the observations $p(z_t|x_t)$ (also called the measurement model). The

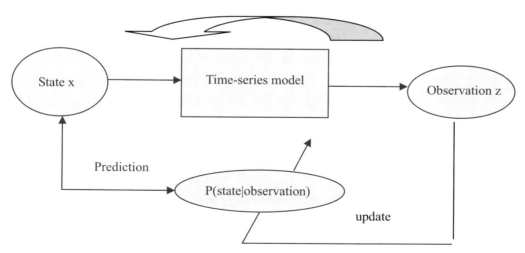

FIGURE 5.2: Block diagram of the state machine for BMIs.

problem is completely defined by specifying the prior distribution $p(x_0)$. The posterior distribution $p(x_{0:t}|u_{1:t}, z_{1:t})$ constitutes the complete solution to the sequential estimation problem, however many times we are interested in one of the marginals, such as the *filtering density* $p(x_t|u_{1:t}, z_{1:t})$, because one does not need to keep track of all the model parameters, and still can answer many important modeling questions.

According to (5.6), there are two models required to analyze and infer the state of a dynamical system: the system model, which describes the evolution of the state with time, and the continuous observation measurement model, which relates the noisy measurements to the state. There are two stages to adapt the filtering density: prediction and update. The prediction stage uses the system model $p(x_{t+1}|x_t)$ to propagate into the future the posterior probability density of the state given the observation, called the Chapman–Kolmogorov equation $p(x_{t-1}|u_{1:t-1}, z_{1:t-1})$ as follows

$$p(x_t \mid u_{1:t-1}, z_{1:t-1}) = \int p(x_t \mid x_{t-1}) p(x_{t-1} \mid u_{1:t-1}, z_{1:t-1}) dx_{t-1}, \tag{5.7}$$

The update stage applies Bayes rule when new data (u_t, z_t) is observed

$$p(x_t \mid u_{1:t}, z_{1:t}) = \frac{p(z_t \mid x_t, u_t) p(x_t \mid x_{1:t-1}, z_{1:t-1})}{p(u_t \mid u_t, z_{1:t-1})} \tag{5.8}$$

To evaluate (5.6) and (5.7), one needs to still compute the system model from the process noise v_{t-1} as

$$p(x_t \mid x_{t-1}) = \int p(x_t \mid v_{t-1}, x_{t-1}) p(v_{t-1} \mid x_{t-1}) dv_{t-1} = \int \Xi(x_t - v_{t-1} - x_{t-1}) p(v_{t-1}) dv_{t-1} \tag{5.9}$$

where the notation $\Xi(.)$ means that the computation is deterministic, and the conditional probability reduces to $p(v_{t-1})$ because of the assumed independence of the noise. The likelihood density function is determined by the measurement model

$$p(z_t \mid x_t, u_t) = \int \Xi(z_t - \hat{f}(u_t, x_t) - n_t) p(n_t) dn_t \tag{5.10}$$

Although the evidence is estimated by

$$p(z_t \mid z_{1:t-1}, u_t) = \int p(z_t \mid x_t, u_t) p(x_t \mid z_{1:t-1}, u_{1:t-1}) dx_t \tag{5.11}$$

When one closely examines the procedure of computing these expressions, it is clear that the difficulty is one of integrating PDFs. The most popular approach is to assume a Gaussian distribution for the PDFs and proceed with closed form integration, which provides the Kalman filter when the model is linear [17], and its variants (extended Kalman filter (EKF) and the unscented

Kalman filter). The most general case is called the particle filter where the system and observation models can be nonlinear and there are no modeling constrains imposed on the posterior density. The particle filter became very popular recently with the availability of fast processors and efficient algorithms, such as Monte Carlo integration, where a set of samples drawn from the posterior distribution of the model parameters is used to approximate the integrals by sums [25]. Alternatively, the sum of Gaussian models can be used, or more principally graphical models, which take advantage of known dependencies in the data to simplify the estimation of the posterior density.

5.3   KALMAN FILTER

Our discussion of generative models will begin with the most basic of the Bayesian approaches: the Kalman filter. The Kalman filter has been applied in BMI experimental paradigms by several groups [7, 16, 27–29]. This approach assumes a linear relationship between hand motion states and neural firing rates (i.e., continuous observations obtained by counting spikes in 100-msec windows), as well as Gaussian noise in the observed firing activity. The Kalman formulation attempts to estimate the state, $\mathbf{x}(t)$, of a linear dynamical system as shown in Figure 5.3. For BMI applications, we define the states as the hand position, velocity, and acceleration, which are governed by a linear dynamical equation as shown in (5.12)

$$\mathbf{x}(t) = [\mathbf{HP}(t) \quad \mathbf{HV}(t) \quad \mathbf{HA}(t)]^{\mathrm{T}} \qquad (5.12)$$

where $\mathbf{HP}$, $\mathbf{HV}$, *and* $\mathbf{HA}$ are the hand position, velocity, and acceleration vectors,[1] respectively. The Kalman formulation consists of a generative model for the data specified by a linear dynamic equation for the state in (5.13)

$$\mathbf{x}(t + 1) = \mathbf{A}\mathbf{x}(t) + \mathbf{u}(t) \qquad (5.13)$$

where $\mathbf{u}(t)$ is assumed to be a zero-mean Gaussian noise term with covariance $\mathbf{U}$. The observation model also called the output mapping (from state to spike trains) for this BMI linear system is simply

$$\mathbf{z}(t) = \mathbf{C}\mathbf{x}(t) + \mathbf{v}(t) \qquad (5.14)$$

where $\mathbf{v}(t)$ is the zero-mean Gaussian measurement noise with covariance $\mathbf{V}$ and $\mathbf{z}$ is a vector consisting of the neuron firing patterns binned in nonoverlapping (100 msec) windows. In this specific formulation, the output-mapping matrix $\mathbf{C}$ has dimension $N \times 9$. Alternatively, we could have

---

[1] The state vector is of dimension $9 + N$; each kinematic variable contains an x, y, and z component plus the dimensionality of the neural ensemble.

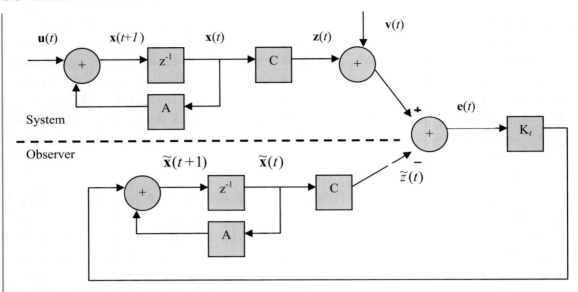

FIGURE 5.3: Kalman filter block diagram.

also included the spike counts of N neurons in the state vector as $\mathbf{f}_1, \ldots, \mathbf{f}_N$. This specific formulation would exploit the fact that the future hand position is not only a function of the current hand position, velocity, and acceleration, but also the current cortical firing patterns. However, this advantage comes at the cost of large training set requirements, because this extended model would contain many more parameters to be optimized. To train the topology given in Figure 5.3, L training samples of $\mathbf{x}(t)$ and $\mathbf{z}(t)$ are utilized, and the model parameters $\mathbf{A}$ and $\mathbf{U}$ are determined using least squares. The optimization problem to be solved is (5.15).

$$\mathbf{A} = \underset{\mathbf{A}}{\arg\min} \sum_{t=1}^{L-1} \|\mathbf{x}(t+1) - \mathbf{A}\mathbf{x}(t)\|^2 \tag{5.15}$$

The solution to this optimization problem is found to be (5.16)

$$\mathbf{A} = \mathbf{X}_1\mathbf{X}_0^T(\mathbf{X}_1\mathbf{X}_1^T)^{-1} \tag{5.16}$$

where the matrices are defined as $\mathbf{X}_0 = \begin{bmatrix} x_1 & \cdots & x_{L-1} \end{bmatrix}$, $\mathbf{X}_1 = \begin{bmatrix} x_2 & \cdots & x_L \end{bmatrix}$. The estimate of the covariance matrix $\mathbf{U}$ can then be obtained using (5.13).

$$\mathbf{U} = (\mathbf{X}_1 - \mathbf{A}\mathbf{X}_0)(\mathbf{X}_1 - \mathbf{A}\mathbf{X}_0)^T /(L-1) \tag{5.17}$$

Once the system parameters are determined using least squares on the training data, the model obtained (**A**, **C**, **U**) can be used in the Kalman filter to generate estimates of the hand positions from neuronal firing measurements. Essentially, this model assumes a linear dynamical relationship between current and future trajectory states.

The Kalman filter is an adaptive state estimator (observer) where the observer gain is optimized to minimize the state estimation error variance. In real-time operation, the Kalman gain matrix **K** (5.19), is updated using the projection of the error covariance in (5.18) and the error covariance update in (5.21). During model testing, the Kalman gain correction is a powerful method for decreasing estimation error. The state in (5.20) is updated by adjusting the current state value by the error multiplied with the Kalman gain.

$$\mathbf{P}^-(t + 1) = \mathbf{A}\mathbf{P}(t)\mathbf{A}^\mathrm{T} + \mathbf{U} \tag{5.18}$$

$$\mathbf{K}(t + 1) = \mathbf{P}^-(t + 1)\mathbf{C}^\mathrm{T}(\mathbf{C}\mathbf{P}^-(t + 1)\mathbf{C}^\mathrm{T})^{-1} \tag{5.19}$$

$$\tilde{\mathbf{x}}(t + 1) = \mathbf{A}\tilde{\mathbf{x}}(t) + \mathbf{K}(t + 1)(\mathbf{Z}(t + 1) - \mathbf{C}\mathbf{A}\tilde{\mathbf{x}}(t)) \tag{5.20}$$

$$\mathbf{P}(t + 1) = (\mathbf{I} - \mathbf{K}(t + 1)\mathbf{C})\mathbf{P}^-(t + 1) \tag{5.21}$$

where the notation $P^-(t + 1)$ means an intermediate (prior) value of P at time $t + 1$. Using the propagation equations above, the Kalman filter approach provides a recursive and on-line estimation of hand kinematics from the firing rate, which is more realistic than the traditional linear filtering techniques and potentially better. The testing outputs for the Kalman BMI are presented in Figure 5.4. The Kalman filter performs better than the linear filter in peak accuracy (CC = 0.78 ± 0.20) but suffers in

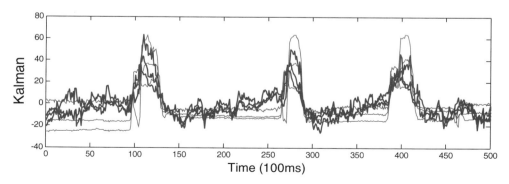

FIGURE 5.4: Testing performance for a Kalman filter for a reaching task. Here, the red curves are the desired x, y, and z coordinates of the hand trajectory, whereas the blue curves are the model outputs.

smoothness because of the number of free parameters (12 073 for a system with 104 neuronal inputs and 9 outputs of 3D position, velocity, and acceleration).

5.4 PARTICLE FILTERS

Although the Kalman filter provides a closed-form decoding procedure for linear Gaussian models, we have to consider the fact that the relationship between neuronal activity and behavior may be nonlinear. Moreover, measured neuronal firing often follows Poisson distributions, and even after binning, there is reason to believe that the Gaussian assumption is too restrictive. The consequences for such a mismatch between model and the real system will be expressed as additional errors in the final position estimates. To cope with this problem, we need to go beyond the linear Gaussian model assumption. In principle, for an arbitrary nonlinear dynamical system with arbitrary known noise distributions, the internal states (**HP**, **HV**, and **HA**) can be still estimated from the measured outputs (neuronal activity) using the sequential estimation framework presented. In the BMI literature, researchers have already implemented the most general of these models called the particle filter [15].

The particle filter framework alleviates the restrictions of the Kalman filter (linearity and Gaussianity) but substantially complicates the computations because, as a result of the generality of the model, there is no closed form solution and therefore the posterior distribution has to be estimated by probing. To help create at the output an estimate of the posterior density, a set of samples drawn from a properly determined density that is estimated at each step, is sent through the system with the present parameters. The peak of this posterior (or another central moment) is considered as the state estimate. Particle filters have also been applied to BMIs [26] where the tuning function has been assumed exponential on linear filtered velocities [32].

In this most general framework, the state and output equations can include nonlinear functions as given in (5.22) and (5.23).

$$x_k = F_k(x_{k-1}, v_{k-1}).$$ (5.22)

$$z_k = H_k(x_k, n_k) .$$ (5.23)

where $F_k : \Re^{n_x} \times \Re^{n_v} \to \Re^{n_x}$ and $H_k : \Re^{n_x} \times \Re^{n_n} \to \Re^{n_z}$ are known, possibly *nonlinear*, functions of the state $\mathbf{x}_{k-1}$, $\{v_{k-1}, k \in N\}$, and $\{n_k, k \in N\}$ are both independent and identically distributed (i.i.d.) process noise; n_x, n_v, n_z, and n_n are dimensions of the state $\mathbf{x}_k$, noise vector $\mathbf{v}_{k-1}$, the measurement $\mathbf{z}_k$, and measurement noise vector $\mathbf{n}_k$, respectively; and N is the set of natural numbers. Note that $\mathbf{v}_k$ and $\mathbf{n}_k$ are assumed *non-Gaussian* in the particle filter.

To deal with these general distributions in a probabilistic way, the integrals have to be evaluated in a numeric way. Particle filtering is implemented to propagate and update the posterior density of the state $\mathbf{x}_k$ given the measurement $\mathbf{z}_k$ recursively over time. Particle filtering uses Sequential Importance Sampling [33] to discretely approximate the posterior distribution. The key idea is to represent the required posterior density function by a set of random samples, which are called particles, with associated weights, and to estimate the posterior density of the state given measurement based on these samples and weights. This Monte Carlo characterization becomes an equivalent representation to the functional description of the posterior PDF of the states when the number of the samples is large enough, and the particle filter solution will also approach the optimal Bayesian estimation.

Let $\{x_{0:k}^i, w_k^i\}_{i=1}^N$ denote a random measure of the posterior PDF $P(x_{0:k}|z_{1:k})$, where $\{x_{0:k}^i\}_{i=1}^N$ is a set of N states up to time k with associated normalized weights $\{w_k^i\}_{i=1}^N$, $\sum_{i=1}^N w_k^i = 1$. Then, the posterior density at time k can be represented by a discrete weighted approximation,

$$p(x_{0:k} \mid z_{1:k}) \approx \sum_{i=1}^N w_k^i \delta(x_{0:k} - x_{0:k}^i). \tag{5.24}$$

The weights are chosen according to the principle of Importance Sampling, which generates samples easily from a proposal Importance Density defined as $q(x_{0:k}|z_{1:k})$ [34, 35],

$$
\begin{aligned}
w_k^i &\propto \frac{p(x_{0:k}^i \mid z_{1:k})}{q(x_{0:k}^i \mid z_{1:k})} \\
&= \frac{p(z_k \mid x_{0:k}^i, z_{1:k-1}) p(x_k^i \mid x_{0:k-1}^i, z_{1:k-1}) p(x_{0:k-1}^i \mid z_{1:k-1})}{p(z_k \mid z_{1:k-1}) q(x_{0:k}^i \mid z_{1:k})} \\
&= \frac{p(z_k \mid x_k^i) p(x_k^i \mid x_{k-1}^i) p(x_{0:k-1}^i \mid z_{1:k-1})}{p(z_k \mid z_{1:k-1}) q(x_{0:k}^i \mid z_{1:k})} \\
&\propto \frac{p(z_k \mid x_k^i) p(x_k^i \mid x_{k-1}^i) p(x_{0:k-1}^i \mid z_{1:k-1})}{q(x_{0:k}^i \mid z_{1:k})}
\end{aligned}
\tag{5.25}
$$

If the importance density is only dependent on x_{k-1} and z_k, we can discard the path $x_{0:k-1}^i$ and the observation history $z_{1:k}$ to simply modify the weight by

$$
\begin{aligned}
w_k^i &\propto \frac{p(z_k \mid x_k^i) p(x_k^i \mid x_{k-1}^i) p(x_{0:k-1}^i \mid z_{1:k-1})}{q(x_{0:k}^i \mid z_{1:k})} \\
&\propto \frac{p(z_k \mid x_k^i) p(x_k^i \mid x_{k-1}^i) p(x_{0:k-1}^i \mid z_{1:k-1})}{q(x_k^i \mid x_{0:k-1}^i, z_{1:k}) q(x_{0:k-1}^i \mid z_{1:k-1})} \\
&= w_{k-1}^i \frac{p(z_k \mid x_k^i) p(x_k^i \mid x_{k-1}^i)}{q(x_k^i \mid x_{k-1}^i, z_k)}
\end{aligned}
\tag{5.26}
$$

Usually, the importance density $q(x_k^i|x_{k-1}^i, z_k)$ is often chosen as the prior density $p(x_k^i|x_{k-1}^i)$ for convenience. The importance density method requires the generation of new samples from $p(x_k^i|x_{k-1}^i)$. The new sample can be generated by the system model with a process noise sample $\mathbf{v}_{k-1}^i$ generated according to the PDF of $\mathbf{v}_{k-1}$. For the importance density that we choose the posterior filtered density $p(x_k|z_{1:k})$ can be approximated as

$$p(x_k \mid z_{1:k}) \approx \sum_{i=1}^{N} w_k^i \delta(x_k - x_k^i)$$

(5.27)

where

$$w_k^i \propto w_{k-1}^i p(z_k \mid x_k^i)$$

(5.28)

It can be shown that when the number of the samples is very large, (5.27) approaches the true posterior density $p(x_k|z_{1:k})$.

The particle filter based on important sampling displays a phenomenon called degeneracy [36]. All but a few particles have negligible weight after several iterations, which implies a large computational effort to update the particles with a minute contribution in the estimation of the posterior density and results in a loss of diversity in the particle pool. When a significant degeneracy appears, *resampling* is implemented to reduce degeneracy [36]. The basic idea of resampling is to eliminate the particles with small weight and to concentrate on particles with large weights. Resampling is applied at every time index, so that the samples are i.i.d. from the discrete uniform density with $w_{k-1}^i = 1/N$. The weight then changes proportionally given by

$$w_k^i \propto p(z_k \mid x_k^i)$$

(5.29)

The weights given by (5.29) are normalized every time before resampling. The assumptions behind resampling are weak. The state dynamics $f_k(x_{k-1}, v_{k-1})$ and measurement function $h_k(x_k, n_k)$ are required to be known. The realization of the posterior density of the state $\mathbf{x}_k$ given the measurement $\mathbf{z}_k$ is sampled from the process noise distribution $\mathbf{v}_k$ and the prior. Finally, the likelihood function $p(\mathbf{z}_k|\mathbf{x}_k)$ is also required to be available to generate new weights. The new sample $x_k^i \sim p(x_k \mid x_{k-1}^i)$ is generated by setting $x_k = f_k(x_{k-1}, v_{k-1})$ with a process noise sample $v_{k-1}^i \sim p_v(v_{k-1})$, where $p_v(v_{k-1})$ is the PDF of v_{k-1}.

The particle filter can be used as a statistical learning and probabilistic inference technique to infer the hand position of a subject from multielectrode recording of neural activity in motor cortex [26].

The hand movement (position, velocity, and acceleration) can be modeled as the system state $\mathbf{x}_k$. The firing rate of the neurons can be modeled as the observation (measurement) $\mathbf{z}_k$. A non-Gaussian model of "tuning" that characterizes the response of the cell firing $\mathbf{z}_k$ conditioned on hand velocity $\mathbf{x}_k$

is explored by adopting particle filter to estimate the posterior distribution recursively. When there is enough neural recordings, the particle filtering method can generate a more complete coverage of the hand movement space. The algorithm can be described as follows:

Initialization N state samples $\mathbf{x}_0^i$, $i = 1, \ldots, N$

For all the time index $k = 1:T$

Draw new samples from $x_k^i \sim p(x_k \mid x_{k-1}^i)$ by setting $x_k = f_k(x_{k-1}, v_{k-1})$ with a process noise sample $v_{k-1}^i \sim p_v(v_{k-1})$

Calculate weights $w_k^i = p(z_k \mid x_k^i)$

Normalize the weights $w_k^i = p(z_k \mid x_k^i) / \sum_{i=1}^{N} p(z_k \mid x_k^i)$

Resample $[\mathbf{x}_k^i, \mathbf{w}_k^i]$ to reduce the degeneracy

Pick out $x_k^{j^*}$ as the estimation state that maximize the posterior density $p(x_k^j \mid z_k)$.

5.5   HIDDEN MARKOV MODELS

The final generative model discussed in this chapter is the HMM. HMMs are famous for capturing the statistical dependencies in time series under the mild assumption that the past of the data is fully described by the present state. For this important reason, HMMs are the leading technology to recognize speech and have even been applied to modeling open-loop human actions, and analyzing similarity between human control strategies. In past efforts, HMMs were trained on neural data to determine the state transition of animal behavior rather than using the likelihood of particular actions [37, 38]. By using HMMs to find likelihoods of simpler movement actions or "movemes" to construct complex motor actions, an unsupervised way to model neural recordings is possible. Here we demonstrate that HMMs are capable of recognizing neural patterns corresponding to two diverse kinematic states (arm at rest or moving) just by analyzing the spatiotemporal characteristics of neural ensemble modulation (i.e., without using a desired signal).

An HMM is a probabilistic model of the joint probability of a collection of random variables $[O_1, \ldots, O_T, X_1, \ldots, X_T]$ (we are going to substitute O for Z to adhere to the notation in the HMM literature). The O_i variables are either continuous or discrete observations, and the X_i variables are *hidden* and *discrete*. Under an HMM, there are two conditional independence assumptions made about these random variables that make associated algorithms tractable. These independence assumptions are:

1. The tth hidden variable, given the $(t-1)^{\text{st}}$ hidden variable, is independent of previous variables

$$P(X_t \mid X_{t-1}, \ldots, X_1) = P(X_t \mid X_{t-1}) \tag{5.30}$$

2. The tth observation, given the tth hidden variable, is independent of other variables

$$P(O_t \mid X_T, O_T, X_{T-1}, O_{T-1}, \ldots, X_t, O_{t-1}, X_{t-1}, \ldots, X_1, O_1) = P(O_t \mid X_t) \tag{5.31}$$

Although continuous and semicontinuous HMMs have been developed, discrete output HMMs are often preferred in practice because of their relative computational simplicity and reduced sensitivity to initial parameter settings during training [39]. A discrete hidden Markov chain depicted in Figure 5.5 consists of a set of n states, interconnected through probabilistic transitions, and it is completely defined by the triplet, $\lambda = [\mathbf{A}, \mathbf{B}, \pi]$, where $\mathbf{A}$ is the probabilistic $n \times n$ state transition matrix, $\mathbf{B}$ is the $L \times n$ output probability matrix (with L discrete output symbols), and π is the n-length initial state probability distribution vector [39, 40].

With respect to the Kalman filter, the HMMs have discrete hidden state variables (as opposed to continuous). Additionally, the HMM has a multistage stochastic state transition, whereas the Kalman filter follows a *single-step* Markov model, following a known deterministic rule, with the randomness appearing through the additive process noise. A graphical model representation of the two topologies in Figure 5.5 makes this similarity even more apparent.

In contrast to the Gaussian noise model that is used for the Kalman filter, the HMM is similar to the particle filter in that it can represent an arbitrary distribution for the next value of the state variables. The differences between the two models become more apparent in the training because the Kalman filter updates are performed sample by sample, whereas in HMMs batch updates are normally used. For an observation sequence O, we locally maximize $P(O|\lambda)$ (i.e., the probability of the observation sequence O given the model λ) by finding the maximum likelihood estimate using an expectation maximization (EM) algorithm called the Baum–Welch algorithm [40].

Recall the definition of the maximum-likelihood estimation problem. We have a density function $p(O|\lambda)$ that is governed by the set of parameters λ and a data set of size N, supposedly drawn from this distribution, that is, $O = [O_1, \ldots, O_N]$. That is, we assume that these data vectors are i.i.d. with distribution p. Therefore, the resulting density for the samples is

$$p(O|\lambda) = \prod_{i=1}^{N} p(O_i|\lambda) = L(\lambda|O) \tag{5.32}$$

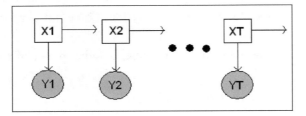 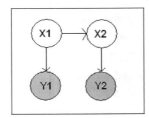

FIGURE 5.5: HMM (left) and Kalman filter (right). Squares denote discrete quantities, circles continuous quantities. White denotes unobservable quantities, and gray observable quantities. Arrows mean dependencies.

This function $L(\lambda \mid O)$ is called the likelihood of the parameters given the data, or just the likelihood function. In the maximum likelihood problem, our goal is to find the λ that maximizes L, that is, find λ^* such that

$$\lambda^* = \operatorname*{argmax}_{\lambda} L(\lambda \mid O) \qquad (5.33)$$

Often, we maximize $\log(L(\lambda \mid O))$ because it is analytically easier. The EM algorithm is particularly useful in situations where one seeks to maximize the likelihood function and the formulation has hidden parameters, which creates uncertainty. Therefore, the observed data do not tell the full story about the problem, and we assume that a complete data set exists $Z = (O, X)$, where X takes into consideration the hidden variables, and specify a joint density function

$$p(z \mid \lambda) = p(O, X \mid \lambda) = p(X \mid O, \lambda) p(O \mid \lambda) \qquad (5.34)$$

From this joint density, we can define the complete data likelihood

$$L(\lambda \mid Z) = L(\lambda \mid O, X) = p(O, X \mid \lambda) \qquad (5.35)$$

Note that $L(\lambda \mid O, X) = h_{\lambda,O}(X)$ where $h_{\lambda,O}(.)$ is some function, λ and O are constants, and X is a random variable (the missing information due to the hidden variable). The original likelihood is referred to as incomplete data likelihood.

The EM algorithm is an iterative algorithm with two steps. Let us call the optimization function we wish to maximize $Q(\lambda, \lambda^{(i-1)})$, where λ stands for the new optimized parameters and $\lambda^{(i-1)}$ the actual parameter set at iteration $i - 1$. First (expectation step), EM finds the expected value of the complete data likelihood $\log p(O, X \mid \lambda)$ with respect to the unknown data X given the observed data O and the current parameter estimates.

$$Q(\lambda, \lambda^{(i-1)}) = E[\log p(O, X \mid \lambda) \mid O, \lambda^{(i-1)}] = \int \log p(O, X \mid \lambda) f(X \mid O, \lambda^{(i-1)}) dX \qquad (5.36)$$

where $f(X \mid O, \lambda^{(i-1)})$ is the marginal distribution of the unobserved data, which is dependent on the observed data and current parameters.

The second step (M step) of the EM algorithm maximizes the expectation computed in the first step, that is,

$$\lambda^{(i)} = \operatorname*{argmax}_{\lambda} Q(\lambda \mid \lambda^{(i-1)}) \qquad (5.37)$$

The beauty of this algorithm is that it has been shown to increase the log likelihood at each step, which means that convergence to a local maximum is guaranteed [40].

The Baum–Welch algorithm is an application of the EM algorithm for HMMs. In an HMM, X_t is a discrete hidden random variable with possible states $[1, \dots, N]$. We further assume that the parameters of the hidden Markov chain are time-independent, that is, $P(X_t|X_{t-1})$ is time homogeneous. Therefore, $P(X_t|X_{t-1})$ can be represented by a stochastic matrix $A[a_{ij}] = p(X_t = j|X_{t-1} = i)$ that does not change over time. The special case $t = 1$ is specified by the initial distribution $\pi_i = p(X_1 = i)$. A particular observation sequence O is described by $O = (O_1 = o_1, \dots, O_T = o_T)$. The probability of a particular observation vector at time t for state j is described by $b_j(o_t) = p(O_t = o_t| X_t = j)$. The complete collection of parameters is represented by $B=[b_j(.)]$.

There are three basic problems in HMM design but here we will only focus on the following two:

Find $p(O|\lambda)$ for some $O=(o_1, \dots, o_T)$. We will use the forward procedure because it is much more efficient than direct evaluation.

Find $\lambda^* = \arg\max_\lambda p(O|\lambda)$ using the Baum–Welch algorithm.

For the forward algorithm, let us define a forward variable $\alpha_t(i)$:

$$\alpha_t(i) = P(O_1,O_2, \dots, O_t, q_t = X_i \mid \lambda) \tag{5.38}$$

which refers to the probability of the partial observation sequence $[O_1, \dots, O_t]$ while being in state X_i at time t, given the model λ [39]. The $\alpha_t(i)$ variables can be computed inductively, and from them, $P(O|\lambda)$ is easily evaluated as explained by Rabiner [39] and Baum et al. [40]. The forward algorithm is defined below:

1. Initialize:

$$\alpha_1(i) = P(O_1, q_1 = X_i \mid \lambda) \tag{5.39}$$

$$\alpha_1(i) = P(O_1 \mid q_1 = X_i, \lambda)P(q_1 = X_i \mid \lambda) \tag{5.40}$$

$$\alpha_1(i) = \pi_i b_i(O_1), i = \{1,\dots,N\} \tag{5.41}$$

2. Induction:

$$\alpha_{t+1}(j) = P(O_1,\dots,O_{t+1}, q_{t+1} = X_j \mid \lambda) \tag{5.42}$$

$$\alpha_{t+1}(j) = \left[\sum_{i=1}^{N} P(O_1,\dots,O_t, q_t = X_i \mid \lambda)P(q_{t+1} = X_j \mid q_t = X_i, \lambda) \right] P(O_{t+1} \mid q_{t+1} = X_j, \lambda)$$

$$\alpha_{t+1}(j) = \left(\sum_{i=1}^{N} \alpha_t(i)a_{ij} \right) b_j(O_{t+1}), t \in \{1,\dots,T\}, j \in \{1,\dots,T\} \tag{5.43}$$

3. Completion:

$$P(O|\lambda) = \sum_{i=1}^{N} \alpha_T(i) \qquad (5.44)$$

The computation of equation (5.42) in the induction step above (step 2) accounts for all possible state transitions from time step t to time step $t + 1$, and the observable at time step $t + 1$. Figure 5.6 below illustrates the induction step graphically.

The second problem is the optimization. To maximize the probability of the observation sequence O, we must estimate the model parameters $(\mathbf{A}, \mathbf{B}, \pi)$ for λ with the iterative Baum–Welch method [40].

Specifically, for the Baum–Welch method, we provide a current estimate of the HMM $\lambda = [\mathbf{A}, \mathbf{B}, \pi]$ and an observation sequence $O = [O_1, \ldots, O_T]$ to produce a new estimate of the HMM given by $\bar{\lambda} = \{\overline{\mathbf{A}}, \overline{\mathbf{B}}, \overline{\pi}\}$, where the elements of the transition matrix $\overline{\mathbf{A}}$ are,

$$\bar{a}_{ij} = \left(\frac{\sum_{t=1}^{T-1} \zeta_t(i,j)}{\sum_{t=1}^{T-1} \gamma_t(i)} \right), \qquad i,j \in \{1, \ldots, N\} \qquad (5.45)$$

Similarly, the elements for the output probability matrix $\overline{\mathbf{B}}$ are given by,

$$\bar{b}_j(k) = \left(\frac{\sum_t \gamma_t(j)(\text{where} \quad \forall O_t = v_k)}{\sum_{t=1}^{T} \gamma_t(j)} \right), \qquad j \in \{1, \ldots, N\}, k \in \{1, \ldots, L\} \qquad (5.46)$$

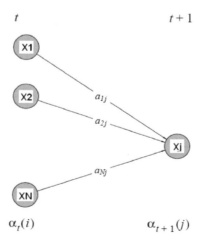

FIGURE 5.6: Forward update.

and finally the $\bar{\pi}$ vector,

$$\bar{\pi} = \gamma_1(i), i \in \{1,...,N\}. \tag{5.47}$$

where,

$$\zeta_t(i,j) = \frac{\alpha_t(i)a_{ij}b_j(O_{t+1})\beta_{t+1}(j)}{P(O|\lambda)} \tag{5.48}$$

and

$$\gamma_t(i) = \sum_{j=1}^{N} \zeta_t(i,j). \tag{5.49}$$

Note that β is the backward variable, which is similar to the forward variable α except that now we propagate the values back from the end of the observation sequence, rather than forward from the beginning of O [40]. First, we defined the backward variables (similar to the forward variables) to,

$$\beta_t(i) = P(O_{t+1},...,O_T | q_t = X_i, \lambda), \tag{5.50}$$

which refers to probability of the partial observation sequence $[O_{t+1}, \ldots, O_T]$ and being in state X_i at time t, given the model λ [39]. Just like the forward variables, the backward variables can be computed inductively.

1. Initialize:

$$\beta_T(i) = 1, i \in \{1,...,N\} \tag{5.51}$$

2. Induction:

$$\beta_t(i) = \sum_{j=1}^{N} a_{ij} b_j(O_{t+1})\beta_{t+1}(j), t \in \{T-1, T-2,...,1\}, i \in \{1,...,N\} \tag{5.52}$$

5.5.1   Application of HMMs in BMIs

The goal of applying HMM to BMIs is different from the previously discussed generative models. With HMMs the hypothesis is that the emerging states correspond to some underlying organization of the local cortical activity that is specific to external events and to their behavioral significance. Specifically, the HMM can be used to recognize a specific pattern (class) of motor neural activity. If we assume that each class of neural data is associated with a given motor behavior, then a

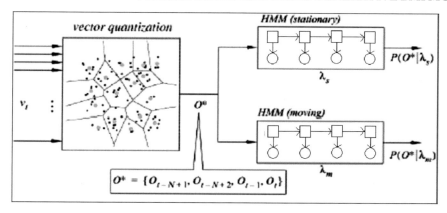

FIGURE 5.7: Stationary/moving classifiers.

forward model can be derived by the techniques discussed in Chapter 3, specifically the gated competitive experts. This allows multiple HMMs to map discrete portions of the neural data to complex trajectories. Consequently, the individual forward models only learn a segment of the trajectory, outperforming single forward model that must generalize over the full trajectory [41].

HMMs were utilized as gates in a gated competitive mixture of linear experts to differentiate and model arm at rest from arm moving in a real-time scenario solely by analysis of the neural recordings. Each one of these two possible outcomes are modeled by an HMM defined by parameters λ_m (movement) and λ_s (stationary), and the goal is to find which one is more likely given that a sequence of neuronal firing rates is observed (Figure 5.7).

There were several difficulties that had to be conquered to apply HMMs to motor BMI data. First, the input data are very high dimensional and discrete, that is 104 neurons of 100-msec binned neural recordings. To decrease the number of parameters of an HMM that would model the posterior density, we created a vector quantization (VQ) preprocessor using the Linde-Buzo-Gray (LBG) algorithm [42] to decrease the training requirements for the HMM and achieve better results. A second method to decrease the number of HMM parameters is to invoke the channel independence assumption and implement an ensemble of single neural-channel HMM chains to form an independently coupled hidden Markov model (ICHMM). Consequently, this classifier takes advantage of the neural firing properties and removes the distortion associated with VQ while jointly improving the classification performance and the subsequent linear prediction of the trajectory.

Vector Quantizing HMM. In this section, we broadly describe the VQ-HMM-based classifier illustrated in Figure 5.7. We demonstrate an experiment in which arm movement is being classified as moving or nonmoving based solely on neural recordings using two HMMs and a winner-take-all strategy. The HMMs act like a switch and allow the neural data belonging to each class to be fed into a feed-forward model to fit it to the respective arm trajectory. To train such a system, we partition the neural recordings into two groups: the first group contains data where the arm appears to be stationary, whereas the second group should contain data where the arm appears to be moving. We use a simple threshold to achieve this grouping: if the instantaneous velocity of the arm is below the noise threshold of the sensor (determined by inspecting the velocity data visually), the corresponding neural data are classified as *stationary*; otherwise, the neural data are classified as *moving* [43]. More specifically, the classifier works as follows:

1)   At time index t, the neural binned vector $\mathbf{v}_t$ of length 104 (equal to number of neurons) is converted into a discrete symbol O_t in preparation for discrete output HMM evaluation.

The multidimensional neural recording must be converted to a sequence of discrete symbols. This process involves vector quantizing the input-space vectors to discrete symbols in order to use discrete-output HMMs. We choose the well-known LBG VQ algorithm [42], which iteratively generates vector codebooks of size $L = 2^m, m \in \{0,1...\}$, and can be stopped at an appropriate level of discretization (represented by m), as determined by the amount of available data. For our example experiment, we varied L from 8 prototype vectors to 256. This optimization seemed to be a good trade off given the 10 000 samples available for training and the 104 dimensional input vector (with each neural bin having about 20 possible bin counts or 20^{104} possible vectors). By optimizing the vector codebook on the neural recordings, we seek to minimize the amount of distortion introduced by the VQ process.

2)   Next, the conditional probabilities $P(O| \lambda_s)$ and $P(O| \lambda_m)$ are evaluated where

$$O = [O_{t-N+1}, O_{t-N+2}, O_{t-1}, O_t], N > 1, \qquad (5.53)$$

and λ_s and λ_m denote HMMs that correspond to the two possible states of the arm (stationary vs. moving).

The vector quantized input generates discrete symbols as input to a left-to-right (or Bakis) HMM chain. Given that we expect the monkey' arm movement to be dependent not only on current neural firings, but also on a recent time history of firings, we train each of the HMM models on observation sequences of length N. During run-time evaluation of $P(O| \lambda_s)$ and $P(O| \lambda_m)$, we use the same value of N as was used during training. Based on other experimental paradigms [44], we varied N from 5 to 10 to correspond to a half of second of data to a second of data. The HMM was trained with the Baum–Welch method on average with five iterations despite our convergence criterion of 0.000001 being met much earlier (because we set the minimum number of iterations to be five). The number of hidden states for the HMM were varied from 2 to 8 as to not exceed the observation sequence length.

3)   Finally, the arm is stationary if,

$$P(O| \lambda_s) > P(O| \lambda_m) \qquad (5.54)$$

and is moving if,

$$P(O| \lambda_m) > P(O| \lambda_s) \qquad (5.55)$$

To explicitly compute $P(O|\lambda)$, we use the practical and efficient forward algorithm [40] described earlier in this section.

The classification decision in (5.53) and (5.54) is too simplistic because it does not optimize for overall classification performance, and does not account for possible desirable performance metrics. For example, it may be very important for an eventual modeling scheme to err on the side of predicting arm motion (i.e., moving class). Therefore, our previous classification decision to include the following classification boundary is modified to :

$$\frac{P(O|\lambda_m)}{P(O|\lambda_s)} = y \qquad (5.56)$$

where y now no longer has to be strictly equal to 1. Note that by varying the value of y, we can essentially tune classification performance to fit our particular requirements for such a classifier. Moreover, optimization of the classifier is now no longer a function of the individual HMM evaluation probabilities, but rather a function of overall classification performance.

In addition, we have determined previously through first- and second-order statistical analysis that some of the 104 recorded neural channels contribute only negligibly to arm movement prediction [41]. Therefore, in the experiments we not only vary the four parameters listed above, but also the subset of neural channels that are used as temporal features in the segmentation process. Table 5.1 lists the seven different subsets of neural channels that were used in our experiments.

Tables 5.2 and 5.3 report experimental results for different combinations of the four parameters and subsets of neural channels. These tables present representative segmentation results for a large number of experiments.

From Table 2, we see that the VQ-HMM can reasonable classify the arm gross dynamics (i.e., moving or not). Unfortunately, even after exhausting many parameter combinations the results were only able to reach a plateau in performance at about 87%. We also note that because a subset of neural channels at the input yielded the best performance, some of the measured neural channels may offer little information about the arm movement. These two observations help to motivate the next section.

TABLE 5.1: Subsets of neural channels	
SUBSET NUMBER	**EXPLANATION**
1	All 104 neural channels
2–5	Different combinations of neural channels with statistically significant correlations between arm movement/nonmovement [3]
6	Neural channels determined to be significant in artificial neural network mappings of neural activity to 3D arm movement
7	51 neural probe channels[a]

[a]The 104-channel neural data through spike sorting of data from 51 neural probes.

In terms of the final trajectory reconstruction, each HMM will be gating a linear model for the particular pieces of the trajectory where it wins the competition (i.e., rest or movement). Each linear model is a multiinput–multioutput finite impulse response (FIR) filter that is trained using normalized least mean square:

$$W_{winner}(n+1) = W_{winner}(n) + \frac{\eta e_{winner}(n)X(n)}{\gamma + \|X(n)\|^2} \qquad (5.57)$$

TABLE 5.2: Results from varying the different parameters and neural subsets					
SUBSET NO.	**STATIONARY %**	**MOVING %**	L	N	n
1	82.1	85.6	8	7	7
4	83.5	87.5	128	10	6
5	84.0	87.5	256	10	6
6	75.6	81.3	256	10	4
7	87.3	86.1	256	10	5

where $\mathbf{w}_{winner}$ is the winning filter's weight vector, $\mathbf{x}(n)$ is the present input, $e_{winner}(n)$ is the error produced from the winning filter, η is the learning rate, and γ is the small positive constant. The data for training is obtained from segments selected by the corresponding HMM (to avoid discontinuities, one second of data is used to fill the FIR memory).

In terms of the final trajectory reconstruction, we see in Figure 5.8 that qualitatively the bimodel system performs well in terms of reaching targets; this is especially evident for the first, second, and the seventh peaks in the trajectory. Overall, prediction performance of the bimodel system is approximately similar to the recurrent neural network (RNN), and superior to the single moving average (MA) model (Wiener filter). The mean of the signal to error ratio (SER) averaged over all dimensions for the single MA model, the RNN, and the bimodal system are -20.3 ± 1.6 (SD), -12.4 ± 16.5, and -15.0 ± 18.8 dB, respectively, whereas the maximum SERs achieved by each model are 10.4, 30.1, and 24.8dB, respectively.

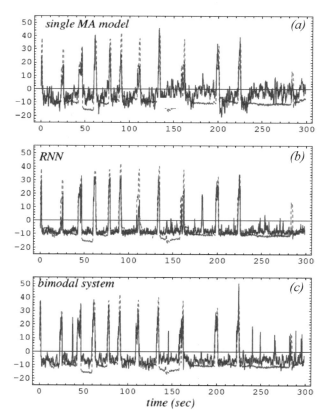

FIGURE 5.8: 3D predicted (blue) and actual arm trajectory (dashed red) for (a) single MA model, (b) RNN, and (c) bimodal system over the test data.

Independently Coupled HMM.   One of the difficulties of the VQ-HMM is the distortion associated with VQ, which limited the performance of the classification. In addition, because different neurons generally have different response properties, dynamic ranges and preferred stimuli, synchronous neural operation is not a good assumption (but rather piecewise stationary). It is difficult, however, to train a HMM with a 104 input vector with the amount of data available in BMI experiments. Therefore, our next implementation invoked an independence assumption among the channels to create an ICHMM.

There is some evidence for this assumption. The literature explains that different neurons in the brain modulate independently from other neurons [45] during the control of movement. Specifically, during movement, different muscles may activate for synchronized directions and velocities, yet, are controlled by independent neural masses or clusters in the motor cortex [2, 12, 45]. Conversely, within the neural clusters themselves, temporal dependencies (and coactivations) have been shown to exist [45]. Therefore, for our HMM classifier, we make the assumption that enough neurons are sampled from different neural clusters to avoid overlap or dependencies. We can further justify this assumption by looking at the correlation coefficients (CC) between all the neural channels in our data set.

The best CCs (0.59, 0.44, 0.42, 0.36) occurred between only four out of thousands of possible neural pairs, whereas the rest of the neural pairs were a magnitude smaller. We believe this indicates weak dependencies between the neurons in our particular data set. In addition, despite these possible weak underlying dependencies, there is a long history of making such independence assumptions to create models that are tractable or computationally efficient. The factorial hidden Markov model is one example among many [46].

By making an independence assumption between neurons, we can treat each neural channel HMM independently. Therefore the joint probability

$$P(O_T^{(1)}, O_T^{(2)}, \ldots O_T^{(D)}, |\lambda_{\text{full}}) \tag{5.58}$$

becomes the product of the marginals

$$\prod_{i=1}^{D} P(O_T^{(i)} | \lambda^{(i)}) \tag{5.59}$$

of the observation sequences (each length T) for each dth HMM chain λ. Because the marginal probabilities are independently coupled, yet try to model multiple hidden processes, we name this classifier the ICHMM.

By using an ICHMM instead of a fully coupled HMM (FCHMM) (Figure 5.9), the overall complexity reduces from ($O(TN^{2D})$ or $O(TD^2N^2)$) to $O(DTN^2)$ given that each HMM chain has a complexity of $O(TN^2)$. In addition, because we are using a single HMM chain to train on a

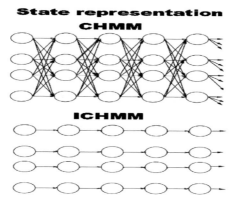

FIGURE 5.9: State representation of ICHMM vs. CHMM.

single neural channel, the number of parameters is greatly reduced and can support the amount of training data. Specifically, the individual HMM chains in the ICHMM contain around 70 parameters for a training set of 10 000 samples as opposed to almost 18 000 parameters necessary for a comparable FCHMM (due to the dependent states).

Each of the individual HMM chains is trained with the Baum–Welch algorithm as before, but now directly from the input data. We compose the likelihood ratio for the decision by

$$l(O) = \frac{\prod_{i=1}^{D} P(O_T^{(i)} | \lambda_M^{(i)})}{\prod_{i=1}^{D} P(O_T^{(i)} | \lambda_r^{(i)})} > \zeta \qquad (5.60)$$

To give a qualitative interpretation of these weak classifiers, we present in Figure 5.10 the probabilistic ratios from 14 single-channel HMM chains (shown between the top and bottom

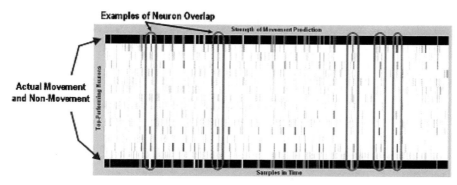

FIGURE 5.10: Example of probabilistic overlap.

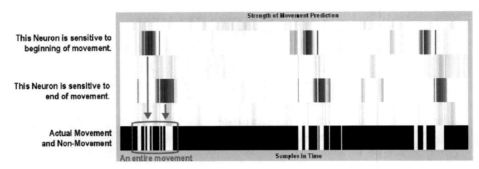

FIGURE 5.11: Zoomed-in example of probabilistic overlap.

movement segmentations) that produced the best classifications individually. Specifically, we present the ratio $\dfrac{P(O^{(i)}|\lambda_m^{(i)})}{P(O^{(i)}|\lambda_r^{(i)})}$ for each neural channel in a grayscale gradient format. The darker bands represent ratios larger than 1 and correspond to a higher probability for the movement class. Lighter bands represent ratios smaller than 1 and correspond to a higher probability for the rest class. The conditional probabilities nearly equal to one another show up as gray bands, indicating that classification for the movement or rest classes is inconclusive.

Overall, Figure 5.10 illustrates that the single channel HMMs can roughly predict movement and rest segments from the neural recordings. In Figure 5.11, they even seem tuned to certain parts of the trajectory like rest-food, food mouth, and mouth-rest. Specifically, we see there are more white bands $P(O|\lambda_r) \gg P(O|\lambda_m)$ during rest segments and darker bands $P(O|\lambda_m) \gg P(O|\lambda_r)$ during movement segments.

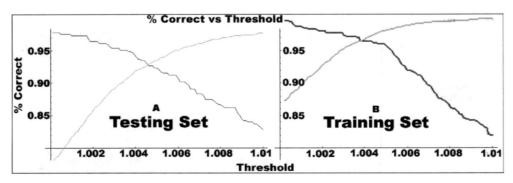

FIGURE 5.12: Training and testing with neural data.

TABLE 5.3: Results compared against different systems on same data

	SINGLE LINEAR MODEL	BIMODEL WITH VQ-HMM	BIMODEL WITH ICHMM
Classification %	–	87%	93%
CC (move)	0.75 ± 0.20	0.79 ± 0.27	0.86 ± 0.11
CC (rest)	0.06 ± 0.22	0.06 ± 0.24	0.03 ± 0.26

Although the standard receiver operator characteristic (ROC) curve can determine the threshold for an appropriate performance (i.e., compromise between number of false alarms and false detections), it fails to convey an explicit relationship between the classes and the threshold when optimizing performance. We use the true positives and true negatives of the respective classes to provide an explicit interclass relationship. For nomenclature simplicity, we label these two curves the likelihood ratio operating characteristic (LROC) curves because they represent the quantities in the likelihood ratio. The "optimal" operating point on the LROC occurs when the two curves of the classes intersect because this intersection represents equal classification for the two classes.

In Figure 5.12, the LROC curves show that the ICHMM is a significant improvement in classification over our previous VQ-HMM classifier. This is evident from the equilibrium point showing that the movement and rest classifications occur around 93% as opposed to 87% in our previous work mentioned in the VQ-HMM section. In Figure 5.12, we also see that the threshold is similar in the training and testing (ζ = 1.0044; ζ = 1.0048, respectively), which show that the threshold can be reliably set from the training data.

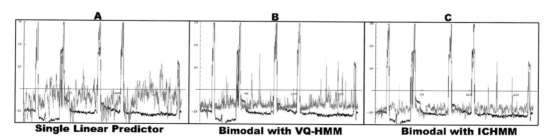

A   **B**   **C**

Single Linear Predictor   **Bimodal with VQ-HMM**   **Bimodal with ICHMM**

FIGURE 5.13: Trajectory results compared.

Figure 5.13 shows the predicted hand trajectories of each modeling approach, superimposed over the desired (actual) arm trajectories for the test data; for simplicity, we only plot the trajectory along the z coordinate. Qualitatively, we can see that the ICHMM performs better than the others in terms of reaching targets. Overall, prediction performance of the ICHMM classifier is slightly better than the VQ-HMM classifier, and superior to the single linear predictor, as evidenced by the average CCs (across the three coordinates) of 0.64, 0.80, and 0.86 for the single linear predictor, the bimodel system with VQ-HMM and the bimodel system with ICHMM.

5.6   SUMMARY

In brain–machine interfaces, the Kalman filter method can provide a rigorous and well-understood framework to model the encoding of hand movement in motor cortex, and for inferring or decoding this movement from the firing rates of a cell population. When the restrictive Gaussian assumptions and linear system model holds, the Kalman filter algorithm provides an elegant analytic optimal solution to the tracking problem. If one assumes that the observation time-series (neural activity) is generated by a linear system, then the tuning can be optimally estimated by a linear filter. The second assumption of Gaussianity of the posterior density of the kinematic stimulus given the neural spiking activities reduces all the richness of the interactions to second order information (mean and the covariance). These two assumptions may be too restrictive for BMI applications and may be overcome with methods such as particle filtering.

Unfortunately, in the BMI application this particular formulation is also faced with problems of parameter estimation. The generative model is required to find the mapping from the low-dimensional kinematic parameter state space to the high-dimensional output space of neuronal firing patterns (100+ dimensions). Estimating model parameters from the collapsed space to the high-dimensional neural can be difficult and yield multiple solutions. For this modeling approach, our use of physiological knowledge in the framework of the model actually complicates the mapping process. As an alternative, one could disregard any knowledge about the system and use a strictly data-driven methodology to build the model. However, if the distributions are not constrained to be Gaussian, but can be described by unbiased consistent means and covariance, the filter can still be optimally derived using a least squared argument.

For the HMMs, the results presented here show that the final prediction performance of the bimodel system using the ICHMM is much better than using the VQ-HMM, and superior to that of a single linear predictor. Overall, the ICHMM produces good results with few parameters. The one caveat to the ICHMM is the reliance on the ζ threshold. Fortunately, this threshold is retrievable from the training set. Interestingly, the ζ threshold can be viewed as global weighting for the two classes in this system. If one frames the ICHMM as mixture of experts (ME) perhaps boosting or bagging could be used to locally weight these simplistic classifiers in future work. The ME generates complex and powerful models by combining simpler models that often map different regions of the input space [47]. With boosting, models are weighted to create a strong ensemble decisions so that a weighted majority "votes" for the appropriate class labeling [47]. This is analogous to what the ICHMM currently does with a global ζ weighting or biasing.

In the next chapter, we will shift the focus from signal processing methods that use estimates of the instantaneous spike rate through binning and concentrate on the spike trains directly. The trade-offs of computing in BMIs directly with spike trains or with rate codes can have an impact on the ability to resolve modulation, correlation, integration, and coincidence in neuronal representations necessary for accurate reconstruction of the sensory or motor function. The implications and significance of the choice of spike traines in BMI approaches will be reviewed next.

REFERENCES

1. Georgopoulos, A.P., A.B. Schwartz, and R.E. Kettner, *Neuronal population coding of movement direction.* Science, 1986. **233**(4771): pp. 1416–1419.

2. Georgopoulos, A., et al., *On the relations between the direction of two-dimensional arm movements and cell discharge in primate motor cortex.* Journal of Neuroscience, 1982. **2**: pp. 1527–1537.

3. Flament, D. and J. Hore, *Relations of motor cortex neural discharge to kinematics of passive and active elbow movements in the monkey.* Journal of Neurophysiology, 1988. **60**(4): pp. 1268–1284.

4. Wessberg, J., et al., *Real-time prediction of hand trajectory by ensembles of cortical neurons in primates.* Nature, 2000. **408**(6810): pp. 361–365.

5. Taylor, D.M., S.I.H. Tillery, and A.B. Schwartz, *Direct cortical control of 3D neuroprosthetic devices.* Science, 2002. **296**(5574): pp. 1829–1832. doi:10.1126/science.1070291

6. Serruya, M.D., et al., *Brain–machine interface: Instant neural control of a movement signal.* Nature, 2002. **416**: pp. 141–142. doi:10.1038/416141

7. Sanchez, J.C., et al. *A comparison between nonlinear mappings and linear state estimation to model the relation from motor cortical neuronal firing to hand movements.* in SAB Workshop on Motor Control in Humans and Robots: on the Interplay of Real Brains and Artificial Devices. 2002. University of Edinburgh, Scotland.

8. Sanchez, J.C., et al. *Input–output mapping performance of linear and nonlinear models for estimating hand trajectories from cortical neuronal firing patterns,* in International Work on Neural Networks for Signal Processing. 2002. Martigny, Switzerland. doi:10.1109/NNSP.2002.1030025

9. Moran, D.W., and A.B. Schwartz, *Motor cortical representation of speed and direction during reaching.* Journal of Neurophysiology, 1999. **82**(5): pp. 2676–2692.

10. Kalaska, J.F., et al., *A comparison of movement direction-related versus load direction-related activity in primate motor cortex, using a two-dimensional reaching task.* Journal of Neuroscience, 1989. **9**(6): pp. 2080–2102.

11. Georgopoulos, A.P., et al., *Mental rotation of the neuronal population vector.* Science, 1989. **243**(4888): pp. 234–236.

12. Thach, W.T., *Correlation of neural discharge with pattern and force of muscular activity, joint position, and direction of intended next movement in motor cortex and cerebellum.* Journal of Neurophysiology, 1978. **41**: pp. 654–676.

13. Scott, S.H. and J.F. Kalaska, *Changes in motor cortex activity during reaching movements with similar hand paths but different arm postures.* Journal of Neurophysiology, 1995. **73**(6): pp. 2563–2567.

14. Todorov, E., *Direct cortical control of muscle activation in voluntary arm movements: A model.* Nature Neuroscience, 2000. **3**(4): pp. 391–398.

15. Gao, Y., et al. *A quantitative comparison of linear and non-linear models of motor cortical activity for the encoding and decoding of arm motions*, in the 1st International IEEE EMBS Conference on Neural Engineering. 2003. Capri, Italy.

16. Wu, W., et al. *Inferring hand motion from multi-cell recordings in motor cortex using a Kalman filter.* in SAB Workshop on Motor Control in Humans and Robots: on the Interplay of Real Brains and Artificial Devices. 2002. University of Edinburgh, Scotland.

17. Kalman, R.E., *A new approach to linear filtering and prediction problems.* Transactions of the ASME-Journal of Basic Engineering, 1960. **82**(Series D): pp. 35–45.

18. Andrieu, C., et al., *Particle methods for change detection, system identification, and control.* Proceedings of the IEEE, 2004. **92**(3): pp. 423–438. doi:10.1109/JPROC.2003.823142

19. Jordan, M.I., Learning in Graphical Models. 1998, New York: MIT Press.

20. Rabiner, L.R. and B.H. Juang, Fundamentals of Speech Recognition. Prentice-Hall Signal Processing Series. 1993, Englewood Cliffs, NJ: PTR Prentice-Hall.

21. Mussa-Ivaldi, F.A., *Modular features of motor control and learning.* Current Opinion in Neurobiology, 1999. **9**: pp. 713–717. doi:10.1016/S0959-4388(99)00029-X

22. Dayan, P. and L. Abbott, Theoretical Neuroscience: Computational and Mathematical Modeling of Neural Systems. 2001, Cambridge, MA: MIT Press.

23. Scott, S.H., *Population vectors and motor cortex: neural coding or epiphenomenon.* Nature, 2000. **3**(4): pp. 307–308.

24. Moran, D.W. and A.B. Schwartz, *Motor cortical activity during drawing movements: population representation during spiral tracing.* Journal of Neurophysiology, 1999. **82**(5): pp. 2693–2704.

25. Chen, M.-H., Q.-M. Shao, and J.G. Ibrahim, Monte Carlo Methods in Bayesian Computation. Springer Series in Statistics. 2000, New York: Springer.

26. Brockwell, A.E., A.L. Rojas, and R.E. Kaas, *Recursive Bayesian decoding of motor cortical signals by particle filtering.* Journal of Neurophysiology, 2003. **91**: pp. 1899–1907. doi:10.1152/jn.00438.2003

27. Erdogmus, D., J.C. Sanchez, and J.C. Principe. *Modified Kalman filter based method for train-ing state-recurrent multilayer perceptrons*. in International Work on Neural Networks for Signal Processing. 2002. Martigny, Switzerland. doi:10.1109/NNSP.2002.1030033

28. Wu, W., M.J. Black, D. Munford, Y. Gao, E. Bienenstock, and J.P. Donoghue, *Modeling and decoding motor cortical activity using a switching Kalman filter*. IEEE Transactions on Biomedical Engineering, 2004. **51**: pp. 933–942. doi:10.1109/TBME.2004.826666

29. Wu, W., et al., *Bayesian population decoding of motor cortical activity using a Kalman filter*. Neural Computation, 2005. **18**: pp. 80–118. doi:10.1162/089976606774841585

30. Sorenson, H.W., Kalman Filtering: Theory and Application. 1985, New York: IEEE Press.

31. Wan, E.A. and R.v.d. Merwe, *The Unscented Kalman Filter*, in Kalman Filtering and Neural Networks, S. Haykin, ed. 2001, New York: Wiley Publishing.

32. Schwartz, A.B., D.M. Taylor, and S.I.H. Tillery, *Extraction algorithms for cortical control of arm prosthetics*. Current Opinion in Neurobiology, 2001. **11**(6): pp. 701–708. doi:10.1016/S0959-4388(01)00272-0

33. Liu, J.S. and R. Chen, *Sequential Monte Carlo methods for dynamic systems*. Journal of the Ameri-can Statistical Association, 1998. **93**: pp. 1032–1044. doi:10.2307/2669847

34. Doucet, A., S. Godsill, and C. Andrieu, *On sequential Monte Carlo methods for Bayesian filtering*. Statistical Computation, 2000. **10**: pp. 197–208.

35. Doucet, A., Monte Carlo Methods for Bayesian Estimation of Hidden Markov Models: Ap-plication to Radiation Signals. 1997, Orsay, France: University Paris-Sud.

36. Arulampalam, M.S., et al., *A tutorial on particle filters for online nonlinear/non-Gaussian Bayesian Tracking*. IEEE Transactions on Signal Processing, 2002. **50**(2): pp. 174–188. doi:10.1109/78.978374

37. Andersen, R.A., et al., *Multimodal representation of space in the posterior parietal cortex and its use in planning movements*. Annual Review of Neuroscience, 1997. **20**: pp. 303–330. doi:10.1146/annurev.neuro.20.1.303

38. Shenoy, K.V., et al., *Neural prosthetic control signals from plan activity*. NeuroReport, 2003. **14**: pp. 591–597. doi:10.1097/00001756-200303240-00013

39. Rabiner, L.R., *A tutorial on hidden markov models and selected applications in speech recognition*. Proceedings of the IEEE, 1989. **77**(2): pp. 257–286. doi:10.1109/5.18626

40. Baum, L.E., et al., *A maximization technique occurring in the statistical analysis of probabilistic functions of Markov chains*. Annals of Mathematical Statistics, 1970. **41**(1): p. 164-171.

41. Kim, S.P., et al., *Divide-and-conquer approach for brain machine interfaces: Nonlinear mixture of competitive linear models*. Neural Networks, 2003. **16**(5–6): pp. 865–871. doi:10.1016/S0893-6080(03)00108-4

42. Linde, Y., A. Buzo, and R.M. Gray, *An algorithm for vector quantizer design.* IEEE Transactions in Communication, 1980. **28**(1): pp. 84–95. doi:10.1109/TCOM.1980.1094577

43. Darmanjian, S., et al. *Bimodel brain–machine interface for motor control of robotic prosthetic*, in IEEE International Conference on Intelligent Robots and Systems. 2003. Las Vegas, NV.

44. Nicolelis, M.A., et al., *Simultaneous encoding of tactile information by three primate cortical areas.* Nature Neuroscience, 1998. **1**(7): pp. 621–630.

45. Freeman, W.J., Mass Action in the Nervous System: Examination of the Neurophysiological Basis of Adaptive Behavior Through EEG. 1975, New York: Academic Press.

46. Ghahramani, Z. and M.I. Jordan, *Factorial hidden Markov models.* Machine Learning, 1997. **29**: pp. 245–275.

47. Jacobs, R.A., et al., *Adaptive mixtures of local experts.* Neural Computation, 1991. **3**: pp. 79–87.

CHAPTER 6

Adaptive Algorithms for Point Processes

Additional Contributor: Yiwen Wang

In this chapter, we review the design of adaptive filters for *point processes* under the Gaussian assumption, and then introduce a Monte Carlo sequential estimation, to probabilistically reconstruct the state from discrete (spiking) observation events. In the previous chapter, we covered a similar topic but the observations were assumed continuous random variables (i.e., firing rates). On the other hand, here they are point processes, that is, 0/1 signals that contain the information solely in their time structure. So the fundamental question is how to adapt the Bayesian sequential estimation models described in Chapter 5 to point processes. The answer is provided by working with the probability of neural firing (which is a continuous random variable). We will use the well-accepted Poisson model of spike generation introduced in Chapter 2, but this time making the firing rate dependent on the system state.

Methods for analyzing spike trains have been applied primarily to understand how neurons encode information [1]. In motor control BMIs, the problem is actually the reverse, where a process called *decoding* [2] identifies how a spike train in motor cortex can explain the movement of a limb. However, the primary methodologies are still inspired by the *encoding* methods. For example, the population vector method of Georgopoulos [3] is a generative model of the spike activity based on the tuning curve concept (preferential firing for a given hand position/speed) that has been extensively utilized in encoding methods. In BMIs, the population vector technique has been championed by Schwartz et al. [4]. All the encoding methods effectively model the PDF of the spike firings. An alternative methodology that effectively bypasses this requirement is the use of maximum likelihood methods, assuming a specific PDF. In neuroscience, the Poisson distribution assumption is very common because it has been validated in numerous experimental setups, but it cannot account for multimodal firing histograms that are often found in neurons. The Poisson model has been improved with a time varying mean to yield what is called the inhomogeneous Poisson model.

A general adaptive filtering paradigm for point processes was recently proposed in [5] to reconstruct the hand position from the discrete observation of neural firings. This algorithm modeled the neural spike train as an inhomogeneous Poisson process feeding a kinematic model through a nonlinear tuning function. The point process counterparts of the Kalman filter, recursive least squares, and steepest descent algorithms were derived and recently compared in the decoding of tuning parameters and states from the ensemble neural spiking activity [6]. The point process analogue of the Kalman filter performs the best, because it provides an adjustable step size to update the state, which is estimated from the covariance information. However, this method still assumes that the posterior density of the state vector given the discrete observation is Gaussian distributed, which is rather unlikely. A Monte Carlo sequential estimation algorithm for point processes has been recently proposed to infer the kinematic information directly from neural spike trains [7]. Given the neural spike train, the posterior density of the kinematic stimulus can be estimated at each time step without the Gaussian assumption. The preliminary simulations have shown a better velocity reconstruction from exponentially tuned neural spike trains.

These spike-based BMI methods require preknowledge of the neuron receptive properties, and an essential stationary assumption is used when the receptive field is estimated from a block of data, which may not account for changes in response of the neural ensemble from open to closed loop experiments [8]. All the encoding methods effectively model the PDF (often called the intensity function) of spike firings. PDF estimation is a difficult problem that is seldom attempted because it requires lots of data and stationary conditions. An alternative methodology that effectively bypasses this requirement is the use of maximum likelihood methods, assuming a specific PDF. Unfortunately, the extension of these methods to multineuron spike trains is still based on the assumption of spike independence, which does not apply when neurons are part of neural assemblies.

6.1 ADAPTIVE FILTERING FOR POINT PROCESSES WITH A GAUSSIAN ASSUMPTION

One can model a point process by using a Bayesian approach to estimate the system state by evaluating the posterior density of the state given the discrete observation [6]. This framework provides a nonlinear time-series probabilistic model between the state and the spiking event [10].

Given an observation interval $(0, T]$, the number $N(t)$ of events (e.g., spikes) can be modeled as a stochastic inhomogeneous Poisson process characterized by its conditional intensity function $\lambda(t \mid \mathbf{x}(t), \mathbf{\theta}(t), \mathbf{Z}(t))$, that is, the instantaneous rate of events, defined as

$$\lambda(t \mid \mathbf{x}(t), \mathbf{\theta}(t), \mathbf{Z}(t)) = \lim_{\Delta t \to 0} \frac{\Pr(N(t + \Delta t) - N(t) = 1 \mid \mathbf{x}(t), \mathbf{\theta}(t), \mathbf{Z}(t))}{\Delta t} \tag{6.1}$$

where $\mathbf{x}(t)$ is the system state, $\mathbf{\theta}(t)$ is the parameter of the adaptive filter, and $\mathbf{Z}(t)$ is the history of all the states, parameters, and the discrete observations up to time t. The relationship between the single-parameter Poisson process λ, the state $\mathbf{x}(t)$, and the parameter $\mathbf{\theta}(t)$ is a nonlinear model represented by

$$\lambda(t \mid \mathbf{x}(t), \mathbf{\theta}(t)) = f(\mathbf{x}(t), \mathbf{\theta}(t)) \tag{6.2}$$

The nonlinear function $f(\cdot)$ is assumed to be known or specified according to the application. Let us consider hereafter the parameter $\boldsymbol{\theta}(t)$ as part of the state vector $\mathbf{x}(t)$. Given a binary observation event ΔN_k over the time interval $(t_{k-1}, t_k]$, the posterior density of the whole state vector $\mathbf{x}(t)$ at time t_k can be represented by Bayes' rule as

$$p(\mathbf{x}_k \mid \Delta N_k, \mathbf{Z}_k) = \frac{p(\Delta N_k \mid \mathbf{x}_k, \mathbf{Z}_k) p(\mathbf{x}_k \mid \mathbf{Z}_k)}{p(\Delta N_k \mid \mathbf{Z}_k)} \tag{6.3}$$

where $p(\Delta N_k \mid \mathbf{x}_k, \mathbf{Z}_k)$ is the probability of observing spikes in the interval $(t_{k-1}, t_k]$, considering the Poisson process

$$\Pr(\Delta N_k \mid \mathbf{x}_k, \mathbf{Z}_k) = (\lambda(t_k \mid \mathbf{x}_k, \mathbf{Z}_k)\Delta t)^{\Delta N_k} \exp(-\lambda(t_k \mid \mathbf{x}_k, \mathbf{Z}_k)\Delta t) \tag{6.4}$$

and $p(\mathbf{x}_k \mid \mathbf{Z}_k)$ is the one-step prediction density given by the Chapman–Kolmogorov equation as

$$p(\mathbf{x}_k \mid \mathbf{Z}_k) = \int p(\mathbf{x}_k \mid \mathbf{x}_{k-1}, \mathbf{Z}_k) p(\mathbf{x}_{k-1} \mid \Delta N_{k-1}, \mathbf{Z}_{k-1}) \mathrm{d}\mathbf{x}_{k-1} \tag{6.5}$$

where the state $\mathbf{x}_k$ evolves according to the linear relation

$$\mathbf{x}_k = F_k \mathbf{x}_{k-1} + \eta_k \tag{6.6}$$

F_k establishes the dependence on the previous state and η_k is zero-mean white noise with covariance Q_k. Substituting (6.4) and (6.5) in (6.3), the posterior density of the state $p(\mathbf{x}_k \mid \Delta N_k, \mathbf{Z}_k)$ can be recursively estimated from the previous value based on all the spike observations.

Assuming the posterior density given by (6.3) and the noise term η_k in the state evolution equation (6.6) are Gaussian distributed, the Chapman–Kolmogorov equation (6.5) becomes a convolution of two Gaussian functions, from which the estimation of the state at each time has a closed-form expression given by (see [11] for details)

$$\mathbf{x}_{k|k-1} = F_k \mathbf{x}_{k-1|k-1} \tag{6.7a}$$

$$P_{k|k-1} = F_k P_{k-1|k-1} F_k' + Q_k \tag{6.7b}$$

$$(P_{k|k})^{-1} = (P_{k|k-1})^{-1} + \left[\left[\frac{\partial \log \lambda}{\partial \mathbf{x}_k} \right]' [\lambda \Delta t_k] \left[\frac{\partial \log \lambda}{\partial \mathbf{x}_k} \right] - (\Delta N_k - \lambda \Delta t_k) \frac{\partial^2 \log \lambda}{\partial \mathbf{x}_k \partial \mathbf{x}_k'} \right]_{\mathbf{x}_{k|k-1}} \tag{6.7c}$$

$$\mathbf{x}_{k|k} = \mathbf{x}_{k|k-1} + P_{k|k} \left[\left[\frac{\partial \log \lambda}{\partial \mathbf{x}_k} \right]' (\Delta N_k - \lambda \Delta t_k) \right]_{\mathbf{x}_{k|k-1}} \tag{6.7d}$$

As expected, the Gaussian assumption allows an analytical solution for (6.5) and therefore, for a closed-form solution of (6.3) as (6.7).

Although the above set of equations may seem daunting, they can be interpreted quite easily. First, (6.7a) establishes a prediction for the state based on the previous value. Then, (6.7b) and (6.7c) are used in (6.7d) to correct or refine the previous estimate, after which the recursive calculation is repeated.

6.2   MONTE CARLO SEQUENTIAL ESTIMATION FOR POINT PROCESSES

The Gaussian assumption applied to the posterior distribution in the algorithm just described may not be true in general. Therefore, for the discrete observations case, a nonparametric approach is developed here that poses no constraints on the form of the posterior density.

Suppose at time instant k the previous system state is $\mathbf{x}_{k-1}$. Recall that because the parameter $\boldsymbol{\theta}$ was embedded in the state, all we need is the estimation of the state from the conditional intensity function, because the nonlinear relation $f(\cdot)$ is assumed known. Random state samples are generated using Monte Carlo simulations [12] in the neighborhood of the previous state according to (6.6). Then, weighted Parzen windowing [13] can be used with a Gaussian kernel to estimate the posterior density. Because of the linearity of the integral in the Chapman–Kolmogorov equation and the weighted sum of Gaussians centered at the samples we are still able to evaluate directly from samples the integral. The process is recursively repeated for each time instant propagating the estimate of the posterior density, and the state itself, based on the discrete events over time. Notice that because of the recursive approach, the algorithm not only depends on the previous observation, but also on the whole path of the spike observation events.

Let $\{\mathbf{x}_{0:k}^i, w_k^i\}_{i=1}^{N_S}$ denote a random measure [14] in the posterior density $p(\mathbf{x}_{0:k} \mid N_{1:k})$, where $\{\mathbf{x}_{0:k}^i, i = 1, \ldots, N_S\}$ is the set of all state samples up to time k with associated normalized weights $\{w_k^i, i = 1, \ldots, N_S\}$ and N_S is the number of samples generated at each time index. Then, the posterior density at time k can be approximated by a weighted convolution of the samples with a Gaussian kernel as

$$p(\mathbf{x}_{0:k} \mid N_{1:k}) \approx \sum_{i=1}^{N_S} w_k^i \cdot k(\mathbf{x}_{0:k} - \mathbf{x}_{0:k}^i, \sigma) \tag{6.8}$$

where $N_{1:k}$ is the spike observation events up to time k modeled by an inhomogeneous Poisson process in previous section, and $k(x - \bar{x}, \sigma)$ is the Gaussian kernel in term of x with mean $\bar{x}$ and covariance σ. By generating samples from a proposed density $q(\mathbf{x}_{0:k} \mid N_{1:k})$ according to the principle of importance sampling [15, 16], which usually assumes dependence on $\mathbf{x}_{k-1}$ and N_k only, the weights can be defined by (6.9).

$$w_k^i \propto \frac{p(\mathbf{x}_{0:k}^i \mid N_{1:k})}{q(\mathbf{x}_{0:k}^i \mid N_{1:k})} \tag{6.9}$$

Here, we assume the importance density obeys the properties of Markov chain such that

$$q(\mathbf{x}_{0:k} \mid N_{1:k}) = q(\mathbf{x}_k \mid \mathbf{x}_{0:k-1}, N_{1:k})q(\mathbf{x}_{0:k-1} \mid N_{1:k-1}) = q(\mathbf{x}_k \mid \mathbf{x}_{k-1}, \Delta N_k)q(\mathbf{x}_{0:k-1} \mid N_{1:k-1}) \tag{6.10}$$

At each iteration, the posterior density $p(\mathbf{x}_{0:k} \mid N_{1:k})$ can be estimated from the previous iteration as (6.11):

$$
\begin{aligned}
p(\mathbf{x}_{0:k} \mid N_{1:k}) &= \frac{p(\Delta N_k \mid \mathbf{x}_{0:k}, N_{1:k-1})p(\mathbf{x}_{0:k} \mid N_{1:k-1})}{p(\Delta N_k \mid N_{1:k-1})} \\[2mm]
&= \frac{p(\Delta N_k \mid \mathbf{x}_{0:k}, N_{1:k-1})p(\mathbf{x}_k \mid \mathbf{x}_{0:k-1}, N_{1:k-1})}{p(\Delta N_k \mid N_{1:k-1})} \times p(\mathbf{x}_{0:k-1} \mid N_{1:k-1}) \\[2mm]
&= \frac{p(\Delta N_k \mid \mathbf{x}_k)p(\mathbf{x}_k \mid \mathbf{x}_{k-1})}{p(\Delta N_k \mid N_{1:k-1})} \times p(\mathbf{x}_{0:k-1} \mid N_{1:k-1}) \\[2mm]
&\propto p(\Delta N_k \mid \mathbf{x}_k)p(\mathbf{x}_k \mid \mathbf{x}_{k-1})p(\mathbf{x}_{0:k-1} \mid N_{1:k-1})
\end{aligned} \tag{6.11}
$$

Replacing (6.10) and (6.11) into (6.9), the weight can be updated recursively as (6.12):

$$w_k^i \propto \frac{p(\Delta N_k \mid \mathbf{x}_k^i)p(\mathbf{x}_k^i \mid \mathbf{x}_{k-1}^i)p(\mathbf{x}_{0:k-1}^i \mid N_{1:k-1})}{q(\mathbf{x}_k^i \mid \mathbf{x}_{k-1}^i, \Delta N_k)q(\mathbf{x}_{0:k-1}^i \mid N_{1:k-1})} = \frac{p(\Delta N_k \mid \mathbf{x}_k^i)p(\mathbf{x}_k^i \mid \mathbf{x}_{k-1}^i)}{q(\mathbf{x}_k^i \mid \mathbf{x}_{k-1}^i, \Delta N_k)} w_{k-1}^i \tag{6.12}$$

Usually, the importance density $q(\mathbf{x}_k^i \mid \mathbf{x}_{k-1}^i, \Delta N_k)$ is chosen to be the prior density $p(\mathbf{x}_k^i \mid \mathbf{x}_{k-1}^i)$, requiring the generation of new samples from $p(\mathbf{x}_k^i \mid \mathbf{x}_{k-1}^i)$ by (6.6) as a prediction stage.

Sequential Importance Resampling [17] should also be applied at every time index to avoid degeneration, so that the sample is i.i.d. from the discrete uniform density with weights $w_{k-1}^i = \frac{1}{N_S}$. The weights then change proportionally given by

$$w_k^i \propto p(\Delta N_k \mid \mathbf{x}_k^i) \tag{6.13}$$

where $p(\Delta N_k \mid \mathbf{x}_k^i)$ is defined by (6.4) in this section. Using (6.6), (6.13), and the resampling step, the posterior density of the state $\mathbf{x}_k$ given the whole path of the observed events up to time t_k can be approximated as

$$p(\mathbf{x}_k \mid N_{1:k}) \approx \sum_{i=1}^{N_S} p(\Delta N_k \mid \mathbf{x}_k^i) \cdot k(\mathbf{x}_k - \mathbf{x}_k^i) \qquad (6.14)$$

Equation (6.14) shows that the posterior density of the current state given the observation is modified by the latest probabilistic measurement of observing the spike event $p(\Delta N_k \mid \mathbf{x}_k^i)$, which is the update stage in the adaptive filtering algorithm.

Without a closed form solution for state estimation, we estimate the posterior density of the state given the observed spike event $p(\mathbf{x}_k \mid N_{1:k})$ at every step and apply two methods to get the state estimation $\tilde{\mathbf{x}}_k$. One possible method is maximum likelihood estimation (MLE), which picks out the sample $\mathbf{x}_k^{i^*}$ with maximum posterior density, or alternatively the expectation of the posterior density can be picked as the state. As we smooth the posterior density by convolving with a Gaussian kernel, we can easily obtain the expectation $\tilde{\mathbf{x}}_k$, and its error covariance V_k by a technique called collapse [18]:

$$\tilde{\mathbf{x}}_k = \sum_{i=1}^{N_S} p(\Delta N_k \mid \mathbf{x}_k^i) \cdot \mathbf{x}_k^i \qquad (6.15)$$

$$V_k = \sum_{i=1}^{N_S} p(\Delta N_k \mid \mathbf{x}_k^i) \cdot (\sigma + (\mathbf{x}_k^i - \tilde{\mathbf{x}}_k)(\mathbf{x}_k^i - \tilde{\mathbf{x}}_k)^T) \qquad (6.16)$$

From (6.15) and (6.16), it is evident that with simple computation one can easily estimate the next state. Hence, the expectation by collapse is simple and elegant.

The major drawback of the algorithm is computational complexity because the quality of the solution requires many particles $\{\mathbf{x}_{0:k}^i, i = 1, \ldots, N_S\}$ to approximate the posterior density. Smoothing the particles with kernels as in (6.14) alleviates the problem in particular when collapsing is utilized, but still the computation is much higher than calculating the mean and covariance of the PDF with a Gaussian assumption.

6.3   SIMULATION OF MONTE CARLO SEQUENTIAL ESTIMATION USING SPIKE TRAINS

Next, we will describe how the framework presented in Section 6.2 can be used to determine neuronal receptive fields. In a conceptually simplified motor cortical neural model [19], the one-dimensional velocity can be reconstructed from the neuron spiking events by Monte Carlo sequential estimation algorithm. This algorithm can provide a probabilistic approach to infer the most probable velocity as one of the components of the state. This decoding simulation updates the state estimation simultaneously and applies this estimation to reconstruct the signal, which assumes the interdependence between the encoding and decoding so that the accuracy of the receptive field estimation and the accuracy of the signal reconstruction are reliable. Notice that dealing with a point process is a more complex problem than when firing rates (continuous observations) are used as traditionally considered in particle filtering [20].

Let us first explain how the simulated data was generated. The tuning function of the receptive field that models the relation between the velocity and the firing rate is assumed exponential and given by (6.17)

$$\lambda(t_k) = \exp(\mu + \beta_k v_k) \tag{6.17}$$

where $\exp(\mu)$ is the background firing rate without any movement and β_k is the modulation in firing rate due to the velocity v_k. In practice in the electrophysiology laboratory, this function is unknown. Therefore, an educated guess needs to be made about the functional form, but the exponential function is widely utilized [21].

The desired velocity was generated as a random walk with a noise variance 2.5×10^{-5} at each 1-msec time step, as shown in Figure 6.1.

The background-firing rate $\exp(\mu)$ and the modulation parameter β_k are set as 1 and 3, respectively, for the whole simulation time, 200 sec. A neuron spike is drawn as a Bernoulli random variable with probability $\lambda(t_k)\Delta t$ within each 1-msec time window [22]. The neuron spike train is shown in Figure 6.2.

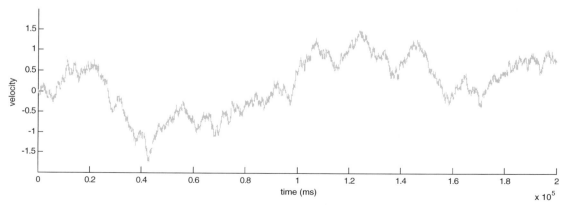

FIGURE 6.1: The desired velocity generated by random walk.

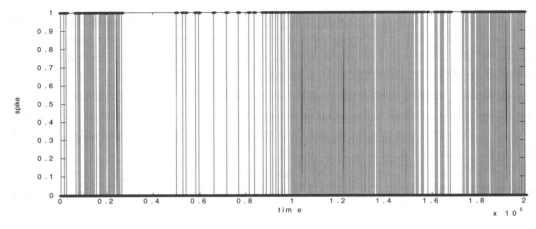

FIGURE 6.2: Simulated neuron spike train generated by an exponential tuning function.

With the exponential tuning function operating on the velocity, we can see that when the velocity is negative, there are few spikes; whereas when the velocity is positive, many spikes appear. The problem lies in obtaining from this spike train the desired velocity of Figure 6.1, assuming the Poisson model of (6.17) and of course using one of the sequential estimation techniques discussed.

To implement the Monte Carlo sequential estimation of the point process, we regard both modulation parameter β_k and velocity v_k as the state $\mathbf{x}_k = [v_k \ \beta_k]^T$. One hundred samples of velocity v_0^i and modulation parameter β_k were initialized, respectively, with a uniform and with a Gaussian distribution. The new samples are generated according to the linear state evolution (6.6), where $F_k = \text{diag}[1 \ 1]$ and Q_k is the covariance matrix of the i.i.d. noise. The kernel size utilized in (6.14) to estimate the maximum of the posterior density (through MLE) was the average spike interval.

To obtain realistic performance assessments of the different models (MLE and collapse), the state estimations $\tilde{v}_k \ \tilde{\beta}_k$ for the duration of the trajectory are drawn 20 different times for different runs of the noise covariance matrices Q_k for state generation. The MSE between the desired trajectory and the model output is shown in Table 6.1 for the adaptive filtering and sequential estimation with both MLE and collapse. In general, if Q_k is too large, one needs many samples to estimate the PDF of the state appropriately and the state trajectory may not be smooth. If it is too small, the reconstructed velocity may get stuck in the same position, whereas the simulated one moves away by a distance much larger than Q_k.

The best velocity reconstruction, shown in the first row of Table 6.1, by both methods is shown in Figure 6.1.

From Figure 6.3 and Table 6.1, we can see that compared with the desired velocity (dash–dotted line) the best velocity reconstruction was achieved by the sequential estimation with collapse (solid black line). It is more sensitive than the adaptive filtering approach involving the Gaussian

MSE DIAG(Q_k)	ADAPTIVE FILTERING OF POINT PROCESS	SEQUENTIAL ESTIMATION	
		MLE	COLLAPSE
[2e-5, 1e-7]	0.04801	0.1199	0.04522
[2e-5, 1e-6]	0.1081	0.1082	0.0489
[1e-5, 1e-6]	0.1076	0.1013	0.0588

TABLE 6.1: Comparison results of all algorithms with different Q_k

assumption (solid gray line). The performance difference can be attributed to a more accurate estimate of the real posterior density because no further assumptions are made. The MLE estimation does not perform well in this simulation for reasons that are not totally clear, but may be related to the multiple maxima of the posterior. As an example in Figure 6.4, the posterior density at time 45.092 sec is shown (blue solid line) with three ripples and is obviously not Gaussian distributed.

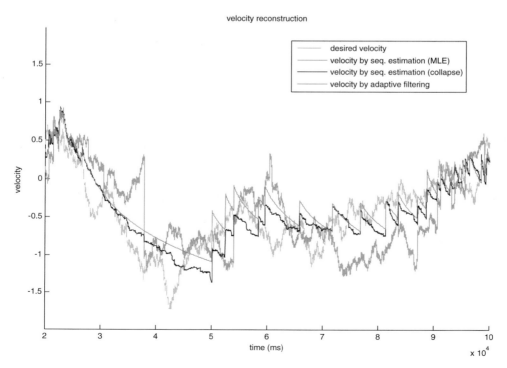

FIGURE 6.3: Velocity reconstruction by different algorithms.

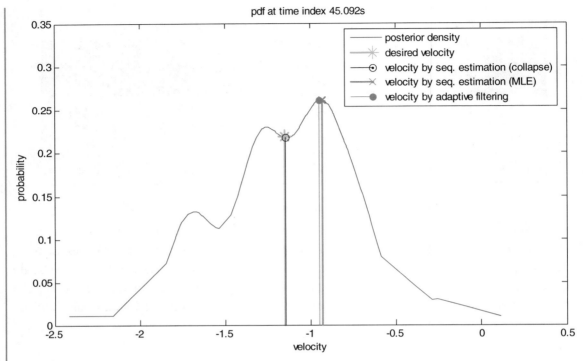

FIGURE 6.4: $p(\Delta N_k \mid v_k^i)$ at time 45.092 sec.

The velocity estimated by the sequential estimation with collapse denoted by the black circle is the closest to the desired velocity (green star).

In summary, the Monte Carlo sequential estimation on point processes shows a good capability to estimate the state from the discrete spiking events.

6.4   ENCODING/DECODING IN MOTOR CONTROL

It is interesting that in black box modeling, the motor BMI is posed as a decoding problem, that is, a transformation from motor neurons to behavior. However, when we use the Bayesian sequential estimation, decoding is insufficient to solve the modeling problem. The full dimension of the difficulty is still partially hidden in the Kalman filter (how to update the state given the observations?), but it becomes crystal clear when we develop the point process approach. To implement *decoding*, it is important to also model how neurons *encode* movement. Therefore, one sees that generative models do in fact require more information about the task and are therefore an opportunity to investigate further neural functionality.

One aspect that underlies the nervous system operation is the different time scales of the neural spike trains (few milliseconds) versus behavior (motor system dynamics are at hundreds of milliseconds to the second scale). Although at the microscale the spike timing is appropriately modeled by a Poisson random process, behavior is exquisitely organized albeit at a much larger time scale. Modeling methods should appropriately exploit this gap in time scales, instead of using simple averaging of spike occurrences in windows, as we do today.

The tuning function is a good example of a model that builds a functional stochastic relationship across time scales. In the estimation of the tuning function, most researchers use an exponential nonlinearity, but for a more realistic assessment, it is necessary to estimate the tuning functions of the neurons in a training set, and also to estimate the delays between the neural firing and the actual movement. This will have to be done in the training data, but once completed—and assuming that the neurons do not change their characteristics—the model is ready to be used for testing (possibly even in other types of movements not found in the training set).

6.4.1 Spike Generation From Neuronal Tuning

Many different functional forms of tuning have been proposed. Most consist of linear projections of the neural modulation on two or three dimensions of kinematic vectors and bias. Moran and Schwartz introduced an exponential velocity and direction tuned motor cortical model [19]. Eden et al have used a Gaussian tuning function for hippocampal pyramidal neurons [6] . These general mathematical models may not be optimal for dealing with the real data because the tuning properties across the ensemble can vary significantly. Moreover, the tuning may change over time. The accuracy of the tuning function estimation can directly affect the modeling in the Bayesian approach and the results of kinematic estimation.

One appropriate methodology is to estimate neural tuning using the training set data obtained in experiments. Marmarelis and Naka [23] developed a statistical method, called white noise analysis, to model the neural responses with stochastic stimuli. This method was improved by Simoncelli et al. [1]. By parametric model identification, the nonlinear property between the neural spikes and the stimuli was directly estimated from data, which is more reliable than just assuming a linear or Gaussian shape. Here we will describe how to use sequential state estimation on a point process algorithm to infer the kinematic vectors from the neural spike train, which is the opposite of sensory representation in spite of the fact that the kinematic vector can be regarded as the outcome of the motor cortex neuromodulation. The tuning function between the kinematic vector and the neural spike train is exactly the observation model between the state and the observed data in our algorithm.

6.4.1.1 Modeling and Assumptions. Recall that in Chapter 2 we show that a tuning function can be modeled as a linear filter coupled to a static nonlinearity followed by a Poisson model, the so-called linear–nonlinear–Poisson (LNP) model (Figure 6.5).

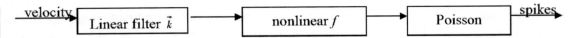

FIGURE 6.5: Block diagram of the linear–nonlinear–Poisson model.

For the BMI, the velocity of the hand movement is measured during a [-300, 500]-msec window with respect to every spike. Because the task is dynamic, a velocity vector at each time step is considered as the input (equal to the size of the space for simplicity). The linear filter projects the velocity vector v into its weight vector k (representing a direction in space), which produces a scalar value that is converted by a nonlinear function f and applied to the Poisson spike-generating model as the instantaneous conditional firing probability for that particular direction in space $p(\text{spike} \mid \vec{k} \cdot \vec{v})$. The filter weights are obtained optimally by least squares $\vec{k} = (E[\vec{v}^T \vec{v}] + \alpha I)^{-1} E_{v|spike}[\vec{v}]$, where $E_{v|spike}$ is the conditional expectation of the velocity data given the spikes, which corresponds to the cross-

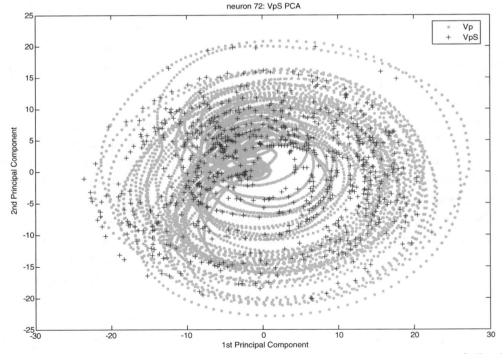

FIGURE 6.6: PCA on spike-triggered velocity vectors compared with the projection of all velocity vectors.

correlation in least squares and for spike trains, but has this form because the binary nature of the data. The parameter α is a regularizing parameter to properly condition the inverse.

This linear filter actually has a geometry interpretation. Figure 6.6 shows the first and the second PCA components of the spike-triggered velocity vectors for neuron 72 (the blue plus symbol in the figure). The velocity vectors for the same segment, which are represented by the green dots, are also projected onto the same PCA component directions. If the neuron has tuning, we expect that the first two components of spike-triggered velocity will be distributed differently from the velocity vectors.

The optimal linear filter actually projects the multidimensional velocity vectors along the direction where they differ the most from the spike-triggered velocity vectors.

For the time interval selected for the spike analysis (i.e., the time interval valid for a Poisson assumption in the collected data), a number is randomly drawn from a normalized uniform distribution (i.e., 0 to 1) and compared with the instantaneous conditional firing probability. If the

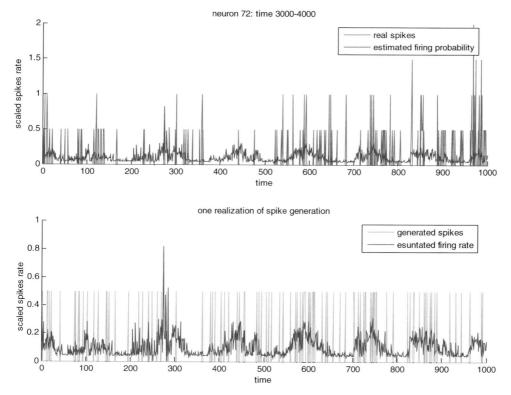

FIGURE 6.7: Spike generation from the estimated tuning function for neuron 72.

number is smaller than the probability, then a spike is generated in this time interval. The tuning function can be shown equivalent to $\lambda_t = f(\vec{k} \cdot \vec{v}_t)$ with $spike_t$ = Poisson (λ_f).

Figure 6.7 shows one realization of a generated spike train using the estimated tuning function for neuron 72. The upper plot depicts spikes scaled by ½ because of the temporal resolution presented in the figure. The blue line is the estimated instantaneous conditional firing probability $p(spike \,|\, \vec{k} \cdot \vec{v})$, which corresponds to the mean λ of the inhomogeneous Poisson process. During most of the duration depicted in the figure, the estimated λ can follow the real spike density change accurately. The bottom plot displays one realization of spikes generated from the estimated λ. The total number of spikes 113 is close to the real number of spike 126 during the time length and the correlation coefficient is between the spike trains smoothed by a Gaussian kernel are shown in Table 6.2 for different number of realizations. Note that the realizations by the Poisson spike generator can be quite different.

6.4.2   Estimating the Nonlinearity in the LNP Model

The nonlinear encoding function *f* for each neuron was estimated using an intuitive nonparametric binning technique. Given the linear filter vector $\vec{k}$, we drew the histogram of all the velocity vectors filtered by $\vec{k}$ and smoothed the histogram by convolving with a Gaussian kernel. The same procedure was repeated to draw the smoothed histogram for the outputs of the spike-triggered velocity vectors filtered by $\vec{k}$. The nonlinear function *f*, which gives the conditional instantaneous firing rate to the Poisson spike-generating model, was then estimated as the ratio of the two smoothed histograms.

Statistically speaking, the kernel smoothed histogram of all the velocity vectors filtered by $\vec{k}$ is an approximation to the marginal PDF $p(\vec{k} \cdot \vec{v})$ of the multidimensional velocity vectors projected

TABLE 6.2: Performance as a function of the number of realizations	
NO. OF REALIZATIONS	**CORRELATION COEFFICIENT**
100	0.89
200	0.94
300	0.96
1000	0.98
10 000	0.99

onto the linear filter direction. The kernel-smoothed histogram of spike-triggered velocity vectors filtered by $\vec{k}$ is an approximation of the joint PDF $p(\text{spike}, \vec{k} \cdot \vec{v})$ projected to the linear filter direction. The ratio actually obeys the Bayesian formulation:

$$p(\text{spike} \mid \vec{k} \cdot \vec{v}) = \frac{p(\text{spike}, \vec{k} \cdot \vec{v})}{p(\vec{k} \cdot \vec{v})} \qquad (6.18)$$

Figure 6.8 depicts the marginal PDF $p(\vec{k} \cdot \vec{v})$, which is the blue line in the top plot, and the joint PDF $p(\text{spike}, \vec{k} \cdot \vec{v})$, which is represented by the red line in the top plot, computed for neuron 110.

The nonlinear estimation for f is shown in the bottom plot of Figure 6.8. The x axis is the range of the filtered multidimensional velocity vectors. The y axis is the corresponding instantaneous conditional probability of spike given the linear filter output. The accuracy of the nonlinear function estimation is not constant across the range, and this may explain the ripples seen at the extremes of the range, so regularization should be utilized. This f is estimated from the real data,

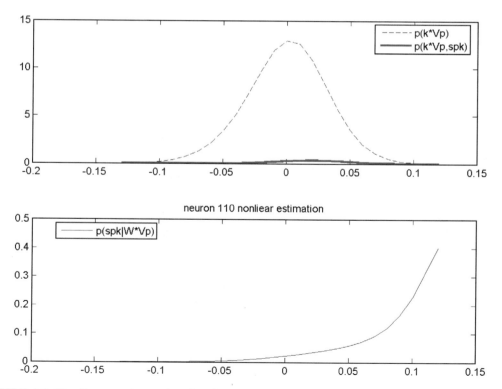

FIGURE 6.8: Nonlinear tuning estimation for neuron 110.

and thus provides more accurate nonlinear properties than just assuming the exponential or Gaussian function. It can actually be implemented as a look up table for the nonlinear tuning function f in testing as

$$p(\text{spike} \mid \vec{k} \cdot \vec{v}_{\text{test}}^{\,t}) = \frac{\sum_{j} k(\vec{k} \cdot \vec{v}_{\text{test}}^{\,t} - \vec{k} \cdot \vec{v}_{\text{spike,training}}^{\,j})}{\sum_{i} k(\vec{k} \cdot \vec{v}_{\text{test}}^{\,t} - \vec{k} \cdot \vec{v}_{\text{training}}^{\,i})} \qquad (6.19)$$

where k is the Gaussian kernel, $\vec{v}_{\text{test}}^{\,t}$ is a possible sample we generate at time t in the test data. $\vec{v}_{\text{spike,training}}^{\,j}$ is one spike-triggered velocity vector sample in the training data, and $\vec{v}_{\text{training}}^{\,i}$ is one velocity vector sample in the training data.

6.4.3 Estimating the Time Delay From the Motor Cortex to Movement

Before implementing the generative BMI based on spikes, one has to align the spike trains with the corresponding hand movements, which is accomplished by estimating the causal time delays due to propagation effects of signals in the motor and peripheral nervous system, as shown in Figure 6.9.

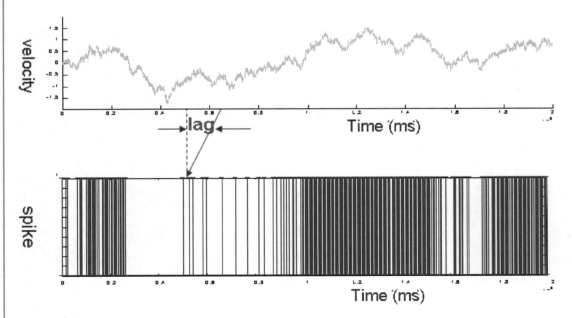

FIGURE 6.9: Illustration of the time delay between neuron spike train (bottom plot) and the kinematics response (upper plot).

To contend with the problem of selecting delays, the mutual information between the spike and the delayed linear filter kinematics vector can be computed as a function of the time lag after a spike arrives. Considering that the response delay ranges from 0 to 500 msec after a neuron fires, the mutual information is determined as a function of lag according to (6.20).

$$I_{(\text{spike};\vec{k}\cdot\vec{x})}(\text{lag}) = \sum_{\vec{k}\cdot\mathbf{x}} p(\vec{k}\cdot\mathbf{x}(\text{lag})) \sum_{\text{spike}=0,1} p(\text{spike}\mid\vec{k}\cdot\mathbf{x}(\text{lag})) \log_2\left(\frac{p(\text{spike}\mid\vec{k}\cdot\mathbf{x}(\text{lag}))}{p(\text{spike})}\right) \qquad (6.20)$$

where to approximate $p(\text{spike}\mid\vec{k}\cdot\mathbf{x}(\text{lag}))$ exactly the same method as the estimation of the nonlinear f we discussed in Section 6.4.1. For all 185 neurons, the mutual information as the function of time delay was obtained from 10 000 continuous samples of kinematics. The time delay with most mutual information was assigned as the best time lag for each neuron.

Figure 6.10 shows the mutual information as the function of time delay from 0 to 500 msec after spikes for five neurons: neurons 72, 77, 80, 99, and 108. The cross on each curve marks the best time lag, which was 110, 170, 170, 130, and 250 msec, respectively. The average best time delay for 185 neurons is 220 msec and was chosen for the subsequent analysis to simplify the implementation.

6.4.4   Point Process Monte Carlo Sequential Estimation Framework for BMIs

The decoding problem for BMIs is shown in Figure 6.11. The spike times from multiple neurons are the multichannel point process observations. The process begins by first windowing the data to determine if a spike is present or not (assign 1 if there is a spike; otherwise, assign 0). The interval should be chosen to be small enough so that only few intervals have more than one spike, and therefore still obey largely the Poisson assumption. One must be careful when selecting the kinematic state (position, velocity, or acceleration) in decoding because it is still not clear which one the neuron actually encodes for. The analysis presented here will consider all three kinematic variables. The velocity is derived as the difference between the current and previous recoded positions, and the acceleration is estimated by first differences from the velocity. For fine timing resolution, all of the kinematics are interpolated and time-synchronized with the neural spike trains.

The whole process can be specified in the following steps:

Step 1: Preprocess and analysis

1.   Generate spike trains from stored spike times.
2.   Synchronize all the kinematics with the spike trains.
3.   Assign the kinematic vector $\vec{\mathbf{x}}$ to be reconstructed.

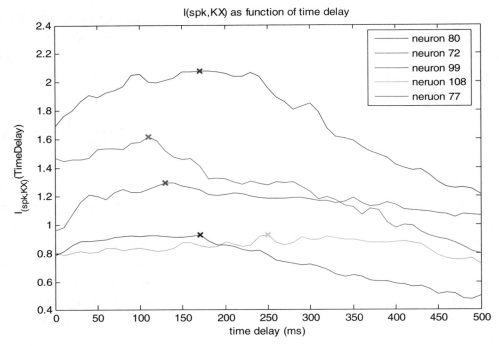

FIGURE 6.10: Mutual information as function of time delay for five neurons.

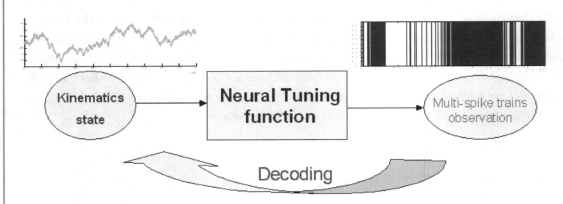

FIGURE 6.11: Schematic of BMI decoding by probabilistic approach.

Step 2: Model estimation

1. Estimate the kinematic dynamics of the system model $F_k = (E[\vec{\mathbf{x}}_{k-1}\vec{\mathbf{x}}_{k-1}^T])^{-1} E[\vec{\mathbf{x}}_{k-1}\vec{\mathbf{x}}_k^T]$
2. For each neuron j, estimate the tuning function

   - Linear model $\vec{k}^j = (E[\vec{\mathbf{x}}^T\vec{\mathbf{x}}])^{-1} E_{\vec{\mathbf{x}}|\text{spike}^j}[\vec{\mathbf{x}}]$
   - Nonlinear function $f^j(\vec{k}^j \cdot \vec{\mathbf{x}}_t) = \dfrac{p(\text{spike}^j, \vec{k}^j \cdot \vec{\mathbf{x}})}{p(\vec{k}^j \cdot \vec{\mathbf{x}})}$
   - Build the inhomogeneous Poisson generator.

Step 3: Monte Carlo sequential kinematics estimation

For each time k, a set of samples for state $\vec{\mathbf{x}}_k^i$ are generated, $i=1{:}N$

1. Predict new state samples $\vec{\mathbf{x}}_k^i = F_k\vec{\mathbf{x}}_k^i + \eta_k$, $i = 1{:}N$
2. For each neuron j,

   - Estimate the conditional firing rate $\lambda_k^{i,j} = f^j(\vec{k}^j \cdot \vec{\mathbf{x}}_k^i)$, $i = 1{:}N$
   - Update the weights $w_k^{i,j} \propto p(\Delta N_k^j \mid \lambda_t^{i,j})$, $i = 1{:}N$

3. Draw the weight for the joint posterior density $W_k^i = \prod_j w_k^{i,j}$, $i=1{:}N$
4. Normalize the weights $W_k^i = \dfrac{W_k^i}{\sum W_k^i}$, $i = 1{:}N$
5. Draw the joint posterior density $p(\vec{\mathbf{x}}_k \mid N_{1:k}) \approx \sum_{i=1}^{N} W_k^i \cdot k(\vec{\mathbf{x}}_k - \vec{\mathbf{x}}_k^i)$
6. Estimate the state $\mathbf{x}_k^*$ from the joint posterior density by MLE or expectation.
7. Resample $\vec{\mathbf{x}}_k^i$ according to the weights W_k^i.

6.4.5 Decoding Results Using Monte Carlo Sequential Estimation

The Monte Carlo sequential estimation framework was tested for the 2D control of a computer cursor. In the preprocessing step we consider a dataset consisting of 185 neurons. Here the state vector is chosen as the instantaneous kinematic vector $\vec{\mathbf{x}} = [\vec{\mathbf{p}}_x\ \vec{\mathbf{v}}_x\ \vec{\mathbf{a}}_x\ \vec{\mathbf{p}}_y\ \vec{\mathbf{v}}_y\ \vec{\mathbf{a}}_y]^T$ to be reconstructed directly from the spike trains. The position, velocity, and acceleration are all included in the vector to allow the possibility of sampling different neurons that represent particular aspects of the kinematics. Picking only the velocity, for example, might provide only partial information between the neural spikes and other kinematics. For each neuron in the ensemble, a small time interval of 10 msec was selected to construct the point process observation sequence. With this interval, 99.62% of windows had a spike count less than 2. For each neuron, 1 was assigned when there was one spike or more appearing during the interval, otherwise 0 was assigned.

After preprocessing the data, the kinematics model F_k can be estimated using the least squares solution. A regularization factor should be used to compute the linear filter weights in the tuning function (1e-4 in this analysis). It should also be noted that this Monte Carlo approach is stochastic, that is, it introduces variation between realizations even with fixed parameters because of the Poisson spike generation model. The most significant parameters include the kernel size σ, the number of particles x_n, and the intensity parameter e_k. For example, in the approximation of the noise distribution $p(\eta)$ approximated by the histogram of $\eta_k = \mathbf{x}_k - F_k\mathbf{x}_{k-1}$, the intensity parameter e_k or number of samples used to approximate the noise distribution can influence computational time and performance of the approximation. The kernel size σ is used to smooth the nonlinearity in the tuning function and in this estimation of the nonlinear function f^j it was assigned 0.02. This kernel size should be chosen carefully to avoid losing the characteristics of the tuning curve and also to minimize noise in the curve.

Figure 6.12 shows one realization of the reconstructed kinematics from all 185 neurons for 1000 samples. The left and right column plots display respectively the reconstructed kinematics for

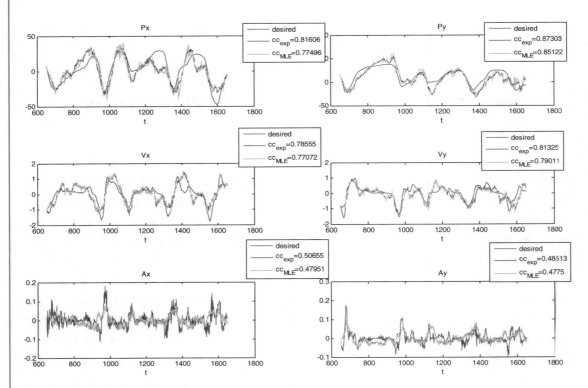

FIGURE 6.12: Reconstructed kinematics for a 2D reaching task.

TABLE 6.3: Average correlation coefficient between the desired kinematics and the reconstructions for 20 realizations

CC	POSITION		VELOCITY		ACCELERATION	
	x	y	x	y	x	y
Expectation	0.7958 ± 0.0276	0.8600 ± 0.0353	0.7274 ± 0.0278	0.7730 ± 0.0155	0.5046 ± 0.0148	0.4873 ± 0.0080
MLE	0.7636 ± 0.0334	0.8327 ± 0.0417	0.7119 ± 0.0292	0.7672 ± 0.0149	0.4872 ± 0.0200	0.4791 ± 0.0115

x axis and y axis. The three rows of plots from top to bottom display respectively the reconstructed position, the velocity, and the acceleration. In each subplot, the red line indicates the desired signal, the blue line indicates the expectation estimation, and green line indicates the MLE estimation. The Monte Carlo approach offers the best reconstruction for the position and the velocity. However, it should be remembered that the reconstructions of the kinematic states will vary among the realizations. The statistical information of correlation coefficients between the desired signal and the estimations for 20 times Monte Carlo runs is given in Table 6.3.

The reconstructed kinematics estimated from the expectation of the joint posterior density performed better than the one from the noise MLE. The position shows the best results as 0.8161 for x and 0.8730 for y than other two kinematic variables. This result may be because velocity and acceleration are derived as differential variables, where the noise in the estimation might be magnified. Another interesting phenomenon is that the y kinematics is reconstructed always better than x, which was the same situation in previous approaches.

To compare the performance with another generative model, Kalman filter algorithm was applied on the same data [24] to predict the hand positions from the binned neural spiking rate. Compared to the approach shown here, the Kalman filter simply assumes both the kinematic dynamic system model and tuning function are linearly related and the posterior density is Gaussian distributed. The average correlation coefficients for reconstructed position and y are 0.62 ± 0.26 and 0.82 ± 0.11, respectively, with sliding window for 40 samples prediction, which is 4 sec long. Because we have different sampling frequency (100 Hz for kinematics rather than 10 Hz), here the average correlation coefficients are calculated with 50% overlapping window. The average correlations for positions x and y are 0.8401 ± 0.0738 and 0.8945 ± 0.0477, respectively, which is better than the Kalman filter results.

6.5   SUMMARY

Although spike trains are very telling of neuronal processing, they are also very removed from the time and macroscopic scales of behavior. Therefore, the spike train methodology begs an answer to the question of how to optimally bridge the time scale of spikes events (milliseconds) with the time scale of behavior (seconds). Most often, the relatively rudimentary method of binning is used, but much of the resolution of the representation is lost, suggesting that better, model-based methodologies need to be developed.

The fundamental problem in spike-based BMI modeling is how to find more effective ways to work directly with spikes for modeling that overcome the difficulties of the conventional approaches. Here we have shown how a Monte Carlo sequential estimation framework could be used as a probabilistic approach to reconstruct the kinematics directly from the multichannel neural spike trains. We investigated an information theoretical tuning depth to evaluate the neuron tuning properties. To describe the functional relationship between neuron firing and movement, a parametric LNP model was utilized. With the knowledge gained from the neuron physiology tuning analysis, a novel signal processing algorithm based on a Monte-Carlo sequential estimation could be applied directly to point processes to convert the decoding role of a brain–machine interface into a problem of state sequential estimation. The Monte Carlo sequential estimation modifies the amplitude of the observed discrete neural spiking events by the probabilistic measurement contained in the posterior density. The Monte Carlo sequential estimation provided a better approximation of the posterior density than point process adaptive filtering with a Gaussian assumption. Compared with the Kalman filter applied to a cursor control task, the preliminary kinematics reconstruction using the Monte Carlo sequential estimation framework showed better correlation coefficients between the desired and estimated trajectory.

However, the signal processing methodologies of spike train modeling are not yet equipped to handle the stochastic and point process nature of the spike occurrence. Many parameters are assumed and need to be estimated with significant complexity. Conventional signal processing methods on random processes work easily with optimal projections; therefore, it would be very useful to create directly from spike trains a metric space that would allow inner product computations because all the available tools of binned methods could be applied immediately.

REFERENCES

1. Simoncelli, E.P., et al., *Characterization of neural responses with stochastic stimuli*. 3rd ed. The New Cognitive Neuroscience. 2004, Cambridge, MA: MIT Press.

2. Paninski, L., et al., *Superlinear population encoding of dynamic hand trajectory in primary motor cortex*. Journal of Neuroscience, 2004. **24**(39): pp. 8551–8561. doi:10.1523/JNEUROSCI.0919-04.2004

3. Georgopoulos, A.P., A.B. Schwartz, and R.E. Kettner, *Neuronal population coding of movement direction*. Science, 1986. **233**(4771): pp. 1416–1419.

4. Schwartz, A.B., D.M. Taylor, and S.I.H. Tillery, *Extraction algorithms for cortical control of arm prosthetics*. Current Opinion in Neurobiology, 2001. **11**(6): pp. 701–708. doi:10.1016/S0959-4388(01)00272-0

5. Brown, G.D., S. Yamada, and T.J. Sejnowski, *Independent component analysis at the neural cocktail party*. Trends in Neurosciences, 2001. **24**(1): pp. 54–63. doi:10.1016/S0166-2236(00)01683-0

6. Eden, U.T., et al., *Dynamic analysis of neural encoding by point process adaptive filtering*. Neural Computation, 2004. **16**: pp. 971–998. doi:10.1162/089976604773135069

7. Wang, Y., A.R.C. Paiva, and J.C. Principe. *A Monte Carlo sequential estimation for point process optimum filtering*, in IJCNN. 2006. Vancouver.

8. Helms Tillery, S.I., D.M. Taylor, and A.B. Schwartz, *Training in cortical control of neuropros-thetic devices improves signal extraction from small neuronal ensembles*. Reviews in the Neurosciences, 2003. **14**: pp. 107–119.

9. Tillery, S.I.H., D.M. Taylor, and A.B. Schwartz, *Training in cortical control of neuroprosthetic devices improves signal extraction from small neuronal ensembles*. Reviews in the Neurosciences, 2003. **14**(1–2): pp. 107–119.

10. Brown, E.N., L.M. Frank, and M.A. Wilson. *Statistical approaches to place field estimation and neuronal ensemble decoding*. Society for Neuroscience Abstracts, 1996. **22**: p. 910.

11. Brown, E.N., et al., *An analysis of neural receptive field plasticity by point process adaptive filtering*. Proceedings of the National Academy of Sciences of the United States of America, 2001. **98**(12): pp. 12 261–12 266. doi:10.1073/pnas.201409398

12. Carpenter, J., P. Clifford, and P. Fearnhead, *Improved particle filter for nonlinear problems*. IEEE Proceedings on Radar, Sonar, and Navigation 1999. **146**: pp. 2–7. doi:10.1049/ip-rsn:19990255

13. Parzen, E., *On the estimation of a probability density function and the mode*. Annals of Mathematical Statistics, 1962. **33**(2): pp. 1065–1076.

14. Arulampalam, M.S., et al., *A tutorial on particle filters for online nonlinear/non-Gaussian Bayesian tracking*. IEEE Transactions on Signal Processing, 2002. **50**(2): pp. 174–188. doi:10.1109/78.978374

15. Bergman, N., Recursive Bayesian Estimation: Navigation and Tracking Applications. 1999, Linkoping, Sweden: Linkoping University.

16. Doucet, A., On Sequential Simulation-Based Methods for Bayesian Filtering. 1998, Cambridge, UK: University of Cambridge.

17. Gordon, N.J., D.J. Salmond, and A.F.M. Smith, *Novel approach to nonlinear/non-Gaussian Bayesian state estimation*. IEEE Proceedings on Radar and Signal Processing, 1993. **140**: pp. 107–113.

18. Wu, W., et al., *Modeling and decoding motor cortical activity using a switching Kalman filter*. IEEE Transactions on Biomedical Engineering, 2004. **51**(6): pp. 933–942. doi:10.1109/TBME.2004.826666

19. Moran, D.W., and A.B. Schwartz, *Motor cortical representation of speed and direction during reaching*. Journal of Neurophysiology, 1999. **82**(5): pp. 2676–2692.

20. Brockwell, A.E., A.L. Rojas, and R.E. Kaas, *Recursive Bayesian decoding of motor cortical signals by particle filtering*. Journal of Neurophysiology, 2003. **91**: pp. 1899–1907. doi:10.1152/jn.00438.2003

21. Trappenberg, T.P., *Fundamentals of Computational Neuroscience*. New York: Oxford University Press, 2002. doi:10.1162/08997660252741149

22. Brown, E.N., et al., *The time-rescaling theorem and its application to neural spike train data analysis*. Neural Computation, 2002. **14**: pp. 325–346.

23. Marmarelis, P.Z., and K.-I. Naka, *White-noise analysis of a neuron chain: An application of the Wiener theory.* Science, 1972. **175**(4027): pp. 1276–1278.

24. Sanchez, J.C., *From cortical neural spike trains to behavior: Modeling and analysis.* 2004, Gainesville, FL: Department of Biomedical Engineering, University of Florida.

• • • •

CHAPTER 7

BMI Systems

Additional Contributors: Erin Patrick, Toshikazu Nishida, David Cheney, Du Chen, Yuan Li, John G. Harris, Karl Gugel, Shalom Darmanjian, and Rizwan Bashirullah

The success of translational neuroprosthetics to aid patients who suffer from neurological disorders hinges on the goal of providing a direct interface for neural rehabilitation, communication, and control over a period of years. Without the assurance that the neural interface will yield reliable recording, preprocessing, and neural encoding/decoding in a small portable (or implantable) package over the patient's lifetime, it is likely that BMIs will be limited to acute clinical interventions. However, the reality of the present systems is far from long life or fully implanted systems.

The present state of BMI technology requires that human or animal subjects be wired to large equipment racks that amplify and process the multidimensional signals collected directly from the brain to control in real time the robotic interface as shown in Figure 7.1. However, the vision for the future is to develop portable BMIs as shown in Figure 7.1c, which incorporates all of the complexity of the laboratory system but in a portable package. Data bandwidths can typically exceed 1Mb/sec when monitoring neural signal activity from multiple electrodes and as a result, many of the systems with wireless links are not fully implantable and require large backpacks [1–7]. To overcome these challenges, several parallel paths have been proposed [7–12] to ultimately achieve a fully implantable wireless neural recording interface. But the truth is that the development of a clinical neural interface requires beyond state-of-the-art electronics, not only DSP algorithms, and models as treated in this book. The purpose of this chapter is to review the basic hardware building blocks that are required for a portable BMI system, establish the bottlenecks, and present developments that are currently underway.

As discussed in Chapter 1, the design space in portable BMIs is controlled by three axes: computation throughput (for real time), power consumption, and size–weight. Obviously, all three depend on the number of recording electrodes, so BMI technology should be specified by *the size–weight, power, and computation per channel*. Each electrode can sense action potentials and LFPs, and creates a digital stream of 300 kbits/sec with conventional analog/digital (A/D) converters

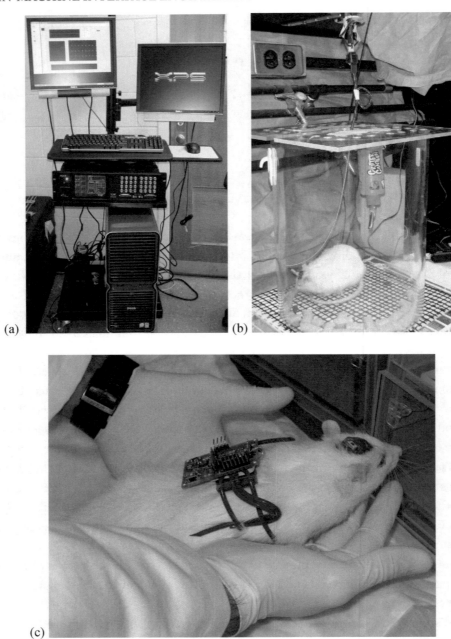

FIGURE 7.1: Experimental BMI systems. (a) Standard recording system used in the electrophysiology laboratory. (b) Tethered recording of a behaving animal. (c) Wireless "backpack" system for BMIs.

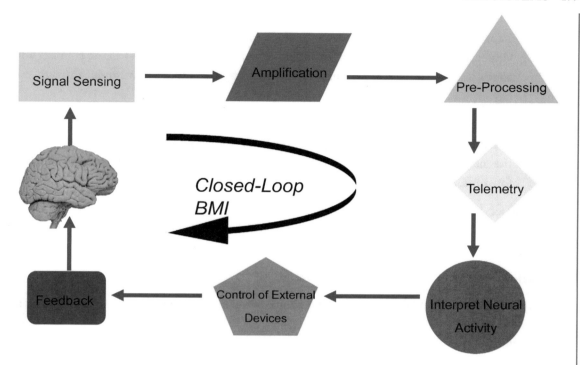

FIGURE 7.2: Hardware block diagram for a closed-loop BMI.

(25 kHz sampling and 12 bits for action potentials, and with 300 Hz with 12 bits for LFPs). Figure 7.2 shows a more detailed block diagram of a closed-loop BMI, where the most important parts are clearly detailed: the electrodes (signal sensing), the amplification, the signal processing (both at the preprocessing and the interpretation), the wireless telemetry, the robotic interface, and the feedback to the user.

We will present two basic approaches for portable BMIs, one short term and another more long term. These approaches will be called first- and second-generation BMI systems.

Generation 1 is a "backpack" approach allowing some mobility to the animal or human patient. Several research groups are working on designs that fall into this category with one or two commercial products in the market. The degrees of freedom (materials, geometries, amplification, A/D, wireless) to accommodate the experimentalist and engineering are enormous, and are visible on the very different engineering solutions proposed and implemented by many research groups [8–23]. Because continuous operation for many hours is a necessity, several modes of operation are possible with the available electrodes and headstages [24, 25]:

Mode 1: Compress the raw neural recordings (action potentials and LFPs) and transmit the full information to a remote processor where it can be recovered with a given precision for visualization and/or BMI model building. The bottleneck in mode 1 is the compression algorithm (i.e., how many channels with which fidelity can be transmitted through the high-speed wireless channel, with a given power budget).

Mode 2: Extract features from raw neural recordings (preprocessing) and transmit only the action potentials or spike counts to a remote station for model building. This is the simplest of the modes, but the issue is to do spike detection and sorting with sufficient precision before binning.

Mode 3: Locally perform the model building and just transmit the commands to the robotic device through a low-bandwidth wireless channel, which requires mode 2. In this case, the wireless channel needs only low bandwidth (send three numbers every 100 msec).

Mode 4: Mix and match among all of the above.

These different possibilities affect the computation and wireless resources, and a flexible hardware solution would be very beneficial.

Generation 2 moves the amplification and preprocessing of neural data into a fully implanted device that transmits the data wirelessly to a base station, where data can be visualized. BMI models are implemented outside the implant and whose outputs control a robotic device through a low-bandwidth wireless channel. Generation 2 systems are at a research stage. The interfaces typically consist of a microelectrode array that can record (motor) or control neural networks chemically (drug delivery) and via stimulation (retinal, cochlear). Because the device is implanted beneath the scalp, it needs to be self-contained; therefore, the electrodes are directly integrated into the electronics, the hardware substrate, and the packaging, and remotely powered, monitored, and controlled via a bidirectional telemetry link [22]. Reports in the literature have provided a wide variety of layouts and physical forms with the entire implant surface area of approximately 100 mm^2 and heights above the skull of no more than several millimeters.

This chapter describes recent progress in each of these areas. Several possible generation 2 architectures are possible and will be briefly reviewed.

7.1   SENSING NEURONAL ACTIVITY: THE ELECTRODES

The ability to quantify neuronal representations has been fueled by advancements in neural recording technology and surgical techniques. The trend since the last century has been to develop acquisition systems with the ability to spatially resolve the activity of large ensembles of single neurons. This goal has been met by advancing electrode design, amplifier design, recording D/A, and surgical technique. For motor control neural interfaces, the functionally representative modulation of activity in cortical columns (i.e., hand, arm, etc., of the homunculus) is the signal of interest. In the rat, cortical columns consist of a dense network of cells with estimates of 100 000 cells/mm3 with pyramidal cells appearing 89% of the time [26]. The ability to resolve sufficient spatial and temporal neuronal representation for the development of BMIs will likely require reliable neural probes that are capable of recording large ensembles of single neurons for long periods of time. The process of extracting signals from the motor, premotor, and parietal cortices of a behaving animal involves the implantation of subdural microwire electrode arrays into the brain tissue (usually layer V) [27]. Current chronic fixed probe technology requires at the surgical stage for the sensor to be placed very close to the neurons of interest for obtaining the highest quality recordings. Moreover, once the probe is placed, the relative distance to the neuron should remain fixed. Dealing with the experimental variability of chronically implantable arrays has been a formidable task because the neural tissue can shift over time with respect to the "fixed" electrode tips, and sensor sites can degrade from immune response and encapsulation. Therefore, the viability of chronic electrodes and recording from single neurons has been limited from months to years. This limitation presents a problem for long-term animal model research and neuroprosthetic development for humans. A variety of approaches have been proposed to improve the electrode tissue interface which include novel electrode material coatings for engineering the glial response and tissue inflamation [28, 29], multisite recording shanks for targeting columnar cortical structure [22, 30], microdrive electrodes for tuning into cellular activity [31], and rapid injectable probes for minimizing implantation injury [32].

Two major types of fixed probes commonly used for neural recording are: 1) conventional wire microelectrodes assembled (most often by hand) from insulated tungsten wires and 2) micromachined electrodes fabricated using integrated circuit microfabrication technologies. Conventional wire microelectrodes are still the "workhorse" of neural recordings because of ease of fabrication and good characteristics in terms of cell yields; however, because they are assembled from discrete components, there is a limit to the electrode array density as well as their functionality. Wire microelectrodes have been extensively used for acute and chronic applications of neural recording [33–35] and have provided the precise firing information of single neurons from cortical and subcortical structures. Typical wire microelectrodes consist of bundles of 50-µm-diameter insulated tungsten wire with sharp electropolished tips. However, the overall bundle size is large for subcutaneous insertion, and the wire spacing is not accurately controlled. For BMIs, the ultimate application of a fully implantable device warrants the need for integration between the amplifiers and electrode arrays. Traditional microwire arrays may not be prepared for this functional integration.

The limitations of mechanically assembled wire electrode arrays can be overcome by using microfabrication and micromachining techniques used in integrated circuits [36, 37]. Moreover, integration of signal processing circuitry onto the substrate of the probe is possible. The recording sites typically consist of exposed metal pads located on rigid shanks that are connected via interconnect traces to output leads or to signal processing circuitry on a monolithic substrate. Some place multiple recording sites along the length of the shank to allow for interrogating cells at varying depths in the neural tissue [36, 37]. An electrode design known as the Utah probe [38–40] uses micromachined square arrays of 100 single contact electrodes bonded to a demultiplexer connector with local amplification. Despite advances in micromachined neural probes, the fixed nature of the probes limit the ability to optimize the recording of the neural signal. If the probe is spatially far from the neuron of interest, the measured action potential is low. The multisite micromachined neural probe provides one approach to resolve this issue. By using a multiplexer to sample the neural signal at different sites at different depths into the neural tissue, a discrete number of locations may be monitored without additional surgery to reposition the probe. An alternative approach is to use moveable probes instead of fixed probes. It is well known that the ability to reposition electrodes after implantation surgery can improve the signal-to-noise ratio (SNR) and yield of neuronal recordings. A variety of microdrives have been proposed that either mechanically or electrically move arrays of electrodes in unison or independently with manual or automated mechanisms. Fabrication of microdrives often requires sophisticated miniature mechanical components and significant skill, time, and cost. For precise targeting of individual units, the probe excursion should traverse 1–2 m in intervals of 5–30 µm. Recently, motorized screw microdrives have been reported to provide a 10-fold increase in yield compared to manual drives [41]. In addition, miniature piezoelectric linear actuators require no gears or screws for electrode movement [31]. Physically, the design challenges also involve specifying small form factors and lightweight components that can fit on the skull of the patient.

Rigid substrate micromachined probes described e may introduce chronic tissue damage due to mechanical forces encountered from strain transferred from the mount to the probes floating in neural tissue [42]. As one way to mitigate this challenge, Rousche et al. [43] designed a flexible substrate microelectrode array, where thin metal electrode sites and subsequent wiring are enclosed between polymers. This flexible electrode array adds needed strain relief, yet cannot be inserted into the brain matter directly. An incision must be created first in order for the electrode to be implanted. Ref. [44] incorporates rigid probes with planar electrode sites on the probe shank hybrid-packaged to a flexible cable that connects electrodes to output ports. This design may be inserted directly into neural tissue, keeping damage to a minimum, and provides greater reliability of the probe assembly.

Here, we report a neural microelectrode array design that leverages the recording properties of conventional microwire electrode arrays with the additional features of precise control of the electrode geometries and flexible micromachined ribbon cable integrated with the rigid probes. The goal is to produce electrode arrays that have high neuronal yield, are highly customizable in terms of geometry/layout, minimize tissue damage, and are easy to mass fabricate.

7.1.1   Microelectrode Design Specifications

An optimal electrode design for BMI chronic in vivo recording requires

- small profile probes that exert the least amount of tissue damage during insertion and chronic recording;
- structurally robust probes that do not buckle during insertion into tissue;
- low probe impedance that is stable during chronic recording;
- recording sites selective to single neuron action potentials;
- adaptability to on-chip processing circuitry.

Using conventional micromachining techniques, one can design small-profile metal traces enclosed between flexible polyimide insulation, making a cable, as shown in Figure 7.3. The actual probes extend from the cable 2 mm and include 20 × 50 µm electrode sites on the tips. The electrode area is chosen for sufficient compromise between signal selectivity and noise performance. The corresponding probe dimensions assure adequate structural integrity according to calculation using the Euler–Bernoulli beam theory. The metal traces and corresponding bond sites can be made to any size specification and spacing distance via photolithography. Therefore, custom application specific

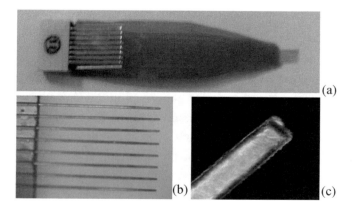

FIGURE 7.3: (a) Flexible substrate microelectrode array with Omnetics connector. (b) Microelectrode array. (c) Probe tip showing insulation along shank and gold plating on tip.

integrated circuits for signal amplification and spike detection may be packaged on the flexible substrate using flip-chip bonding techniques in future designs.

7.1.2 Process Flow

All processing for building the electrode array is performed on the surface of a 100-mm-diameter silicon wafer covered with Kapton tape (which provides adequate adhesion for the subsequent polyimide layers). The polyimide bottom insulation layer (PI 2611; HD Microsystems, Cupertino, CA) is spin deposited and cured to a final thickness of 20 µm. Sputtered nickel, 100 Å, is patterned to define the probe, wiring, and bond pad dimensions. Then, Ni is electrodeposited on the Ni seed to an average thickness of 20 µm via a 10-mA direct current for 4 hours in a nickel sulfamate bath (Nickel S; Technic Inc.). Adhesion promoter (VM9611, HD Microsystems) is next applied followed by three spin coatings of the PI 2611 to achieve the final 20-µm top layer of insulation. Al (1000 Å) is patterned as a hard mask for the subsequent oxygen plasma etch. The etching process includes an O_2 reactive ion etch that removes the polyimide from the top of the bond pads and the probe tips. The remaining polyimide under the probe tips is isotropically etched in a plasma barrel asher. Then the probe-cable assembly is removed from the substrate wafer and primed for parylene-C deposition with a silane adhesion promoter (Acros Organcis, Geel, Belgium). The parylene-C vapor deposition step insulates the shank of the metal probes to a thickness of 2–4 µm. Then the probe ends are manually cut with a blade to expose bare metal for the electrode sites. Finally, the probes are immersed in an electroless gold plating solution (TechniIMGold AT, 600, Technic Inc., Cranston, RI) that covers the electrode sites as well as the bond pad sites with 0.1 µm of gold. An Omnetics connector is then fixed to the bond pads with silver epoxy.

7.1.3 In Vivo Testing

Adult male 250-g Sprague–Dawley rats were used to test the recording performance of the flexible electrode arrays. All procedures have been approved by the University of Florida Institutional Animal Care and Use Committee Board and were performed in the University of Florida McKnight Brain Institute. Before surgery, the rats were anesthetized and the surgical site was thoroughly sterilized. The top of the skull was then exposed by a midsaggital incision from between the eyes, and the landmarks bregma and lambda are located on the skull [45]. The microwire array was implanted to a depth of 1.66 mm into the forelimb region of the primary motor cortex. The electrodes are stereotaxically moved to the appropriate site and lowered to the appropriate depth using a micropositioner (1 mm/h) to minimize distress to the brain tissue (FHC, Bowdowinham, ME) as shown in Figure 7.4. The array was then grounded using a 1/16-in.-diameter stainless steel screw. A low-profile Omnetics connector was used to attach the recording wire.

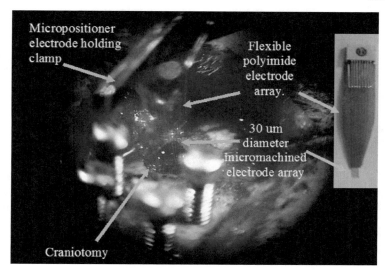

FIGURE 7.4: Implantation of the flexible substrate microwire array.

Extracellular potentials recorded at 12 207 Hz during surgery were analyzed and spike sorted using Spike2 (CED, Cambridge, UK) software package. Recordings were analyzed over a period of 130 sec. To detect and sort neural activity within each channel, an automated waveform matching system within Spike2 was used to construct templates using threshold detection.

Once a set of waveform templates were generated for a data stream, all templates (noise) that did not match characteristic neural depolarization behavior were removed [46–48]. The remaining waveform templates were sorted according to amplitude and shape, and any waveform templates that were significantly similar to each other were combined into a single template. Clustering of waveform variance within templates was verified through PCA. Each waveform template was statistically unique and representative of a distinct neuron within the channel.

Once neuron waveforms were isolated and sorted, peak-to-peak amplitude was evaluated by computing the average waveform of all spikes within the neuron template and measuring the potential difference from the apex of the repolarization peak to the apex of the depolarization peak. The noise floor of each channel was evaluated by computing the root mean square value of a 5-sec period of noise. Using these two values, the SNR for each neuron template was calculated. To ensure proper reporting, all spike waveform templates that possessed peak-to-peak amplitude of magnitude less than three times the value of the noise floor were considered too close to the noise to be reliably and consistently distinguished, and were removed from the study. Values of neural yield, noise floor, amplitude, and SNR are reported for each channel within Table 7.1. Action potential amplitudes as

TABLE 7.1: Neuronal yield for eight channel microelectrode array								
Electrode	1	2	3	4	5	6	7	8
Yield (neurons)	2	2	2	3	6	5	3	4
Noise floor (μV, RMS)	4.1	5.0	5.3	4.4	5.2	3.8	3.7	4.3
Neuron amplitude (μV, PtP)	20.1	23.3	32.6	26.1	114.7	90.4	31.4	45.1
	13.2	15.5	24.7	18.3	56.8	52.3	13.4	29.7
				14.2	34.6	35.7	11.7	21.0
					21.3	21.0		16.0
					18.8	13.8		
					17.4			
SNR (dB)	13.8	13.4	15.8	15.5	26.9	27.6	18.6	20.4
	10.2	9.9	13.4	12.4	20.8	22.8	11.2	16.8
				10.2	16.5	19.5	10.0	13.8
					12.2	14.8		11.4
					11.2	11.2		
					10.5			

large as 115 μV and as small as 13 μV are discriminated by the electrode and recording system. The average RMS noise floor is 4 μV. Figure 7.5 shows recorded data from electrode number 6.

Acute electrophysiological recordings show excellent yield of recordable neurons and SNRs ranging from 10 to 27 dB. The benefits of the microfabrication design allows for tailoring the electrode geometry for neuronal structures of interest. Moreover, high channel count arrays can be constructed by layering the proposed design. The flexible cable additionally provides strain relief from the fixed external connection and can be used to minimize tissue damage in chronic applications. Additional studies on the reliability of the electrode array as a chronic recording device is needed to

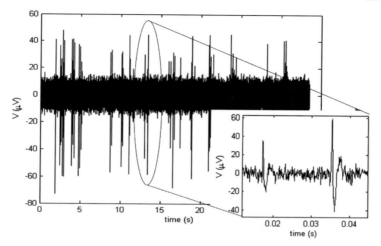

FIGURE 7.5: Data from neural recording in the rat motor cortex at a depth of 1.66 µm during implantation surgery. Inset shows two distinct neurons recorded by single probe.

confirm the biocompatibility of the electrode material. However, because of the adaptability of the fabrication process, other metals such as platinum may be readily incorporated.

7.2 AMPLIFICATION

The next step in the signal processing chain is amplification. Because of the very small signal levels (microvolts) the amplification needs to be done very close to the frontend. The design of the headstage, that is, the connector that links the electrodes and the cables to the amplifier is critical. Impedance matching must be carefully controlled, and most designs use separate ground and reference.

The amplifiers have gains of 40–70 dB, a frequency response between 0.1 Hz and 10 kHz, and are therefore AC coupled. One of the most important requirements is high common mode rejection because of the large electrical artifacts normally present (60 Hz interference and other electrical noises). In rack-mounted instrumentation, the amplifiers are well within the technology envelope and several manufacturers provide excellent instruments (Tucker-Davis Technologies, Plexon, Grass, Biosemi, Axon, Alpha-Omega). The difficulty is in portable designs. We present briefly below an analog amplifier we have developed at the University of Florida (Harris).

7.2.1 Integrated Bioamplifier

The custom amplifier consists of two stages: a low-noise 40-dB preamplifier and a second-stage 40-dB amplifier that were fabricated in the 0.6-µm AMI complementary metal oxide semiconductor

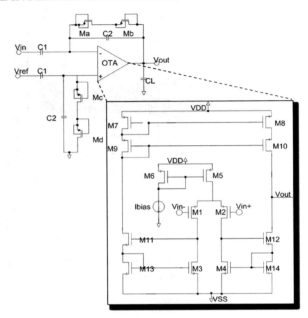

FIGURE 7.6: First-stage amplifier.

(CMOS) process. Figure 7.6 shows the schematic of the preamplifier and the second stage is in Figure 7.7. The amplifier designs are based on the work of Reid Harrison [49].

The first-stage midband gain, A_M, is $-C_1/C_2$, the lower corner frequency is at $\omega_1 \approx 1/(RC_2)$, and the higher corner frequency is at $\omega_2 \approx g_m C_2/(C_L C_1)$, where g_m is the transconductance of the operational transconductance amplifier (OTA) shown in the lower portion of Figure 7.5 [50]. To obtain a low cutoff frequency, a large resistance is needed in the feedback loop provided by two diode-connected transistors, Ma and Mb, acting as "pseudo-resistors" without sacrificing die area.

Noise performance is important in biological preamplifier designs, it is minimized by carefully choosing the width and length of the transistors. However, there is a trade-off between stability and low noise. Because low noise is critical in this application, decreasing the transconductance of M1, M2, M7, M8, and the four transistors of the two Wilson current mirrors can minimize the thermal noise. Flicker noise, which is important in low-frequency applications, is minimized by increasing the device area. Because p type metal oxide semiconductor devices exhibit less flick noise than n type, p type is used in the input pair.

However, the typical neural signal amplitude is on the order of tens to hundreds of microvolts, the output voltage of the preamplifier is still too small for signal processing by an A/D converter. Therefore, a second-stage amplifier is required [51], in which C_1/C_3 determines the midband gain,

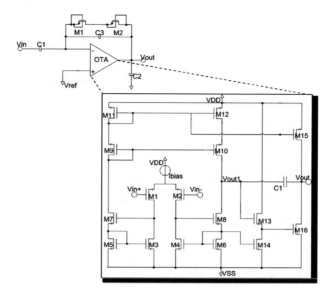

FIGURE 7.7: Second-stage amplifier.

PARAMETER	PREAMPLIFIER	SECOND-STAGE AMPLIFIER
Supply voltage (V)	5	5
Power consumption (µW)	80	~120
Gain (dB)	39.46	~40
Bandwidth (kHz)	5.4	19
Lower cutoff frequency (Hz)	~0.3	<1
Die area (µm²)	91 800	66 400
Input referred noise (µV rms)	5.94	Not available

TABLE 7.2: Amplifier specifications

whereas V_{ref} and diode-connected resistors provide the DC operational bias. The OTA is a typical two-stage OTA with a P-type input differential pair loaded with Wilson current mirrors followed by a class A output stage.

Table 7.2 shows the performance of our current designs per electrode using 0.5-μm technology. Better performance can be obtained with newer technologies. Dies can also be wired bound to the circuit board to minimize the physical size and avoid noise.

7.3   THE PICO SYSTEM

We have developed and tested a data acquisition/preprocessor/wireless system to implement mode 2. We called it the PICO system, and its goal is to acquire neural recordings from the headstage, amplify the signals, perform spike detection (feature extraction/compression), spike sorting, and wirelessly transmit binned data (rate codes) to a remote station. Alternatively, this system can simply amplify the raw neural data, do spike detection with a simple threshold, and send just these windows to a remote system. As is, it already plays an important role in BMIs because it implements a first-generation backpack. However, it does not allow the neurophysiologist to analyze and store the continuous raw voltage traces being collected because of bandwidth limitations (mode 1), nor it is capable of performing the high speed computations of mode 3. This is where a neural signal processor (NSP) (local decoding algorithm implementation) will add capabilities in the future by performing data reduction at the frontend such that the data can be reconstructed in the backend for visualization and archiving purposes. In the particular path described here, the previously reported Pico Recording system [52] is capable of replacing a multiconductor commutator cable used in animal experimental paradigms with a high-bandwidth, low-power wireless interface that allows the subject more mobility and a more natural response to a given experiment. This new wireless interface presents a compact architecture as well as new challenges related to the power consumption of the remote device.

To optimize the trade-off between power consumption and processing power, the low-power, low-noise, high-gain amplifier shown in Section 7.2 was designed in conjunction with off-the-shelf components, specifically MSP430F1611 as the microcontroller and the Nordic nRF2401 wireless transceiver. The processor consumes less than 2 mW of power with 8 million instructions per second of processing and has the added benefit of conserving space by integrating two universal asynchronous receiver/transmitters, three direct memory access channels, and an eight-channel, 12-bit A/D converter all in one small package. In this work, we demonstrate the integration of the low-power amplifier with the Pico system and collect data from a live subject.

The Pico Neural Data Collection System consists of the Pico Remote (battery-powered neural amplifiers, a neural data processor, and a wireless transceiver), the Pico Receiver (a wireless-transceiver to USB converter), and PC-based software (for spike sorting and firing rate quantification). Figure 7.8

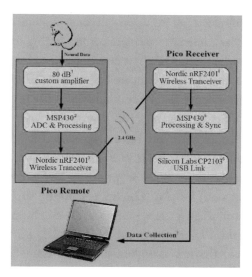

FIGURE 7.8: Block diagram for the PICO wireless neural recording system.

provides a block diagram for signal flow throughout the system. The Pico Remote senses single- and multiunit neuronal activity from 16 chronically implanted microelectrodes from a laboratory rat. The signals are amplified (80 dB) with a custom amplifier1 and fed to one of eight 12-bit A/D channels2 sampling at 20 kHz on the MSP430. The MSP430 has two modes: spike sorting and spike reading. While in the spike sorting mode, raw sampled data from a single probe are sent over the Nordic transceiver for reception by the Pico Receiver. This mode allows the user to analyze in detail the raw recording from the microelectrodes to set thresholds for detecting and sorting spikes (action potentials). In the spike reading mode, the MSP430 executes the spike detection through user defined thresholds and sorting. The firing rate for each neuron is computed locally and sent over the Nordic transceiver for reception by the Pico Receiver.

The Nordic transceiver is capable of up to 250 kbit/sec sustained data transfer of packet data. Data sent over the Nordic transceivers are fed to another MSP430 processor for buffering and formatting and for the eventual pass through to a Silicon Laboratories CP2103 USB bridge to software running on a PC.

The MSP430F1611 processor samples each analog channel at 20 kHz with an on-board 12-bit A/D converter. Depending on the mode in which the processor is functioning, the samples are handled differently. While in the spike sorting mode, no signal processing occurs and the raw samples are sent directly to the transceiver for processing by software running on the PC. The data are sent as 12-bit samples at rate of 20k samples/sec for a data rate of 240k samples/sec. In the spike counting mode, each sample is compared against thresholds that define a spike using well established spike-sorting techniques. If all the threshold conditions are met, a counter is incremented. Each probe can have as many as three rate counters corresponding to three neurons per electrode. Every 100 msec, the bin counts are sent to the wireless transceiver for processing on a PC.

On the receiving side of the RF link, data are collected via a Nordic nRF2401 transceiver on the Pico Receiver. The RF link is controlled by another MSP430F1611 microcontroller and results are passed on to a PC via a USB bridge (CP2103 from Silicon Laboratory). The USB bridge is compliant with the USB 2.0 specification and has a maximum throughput of 12 Mbits/sec.

The Pico Remote is configured with software running on Windows-based PCs. The testing software has three separate applications: Spike Capture, Spike Detection Parameter Initialization, and Binned Spike Data (firing rate) Display. The Spike Capture application is a Matlab graphical user interface that captures and displays raw sample data sent from the Pico Remote. This application does not configure the Pico Remote but is used solely for viewing and storing a single channel's raw sample data for analysis. The Spike Detection Parameter Initialization application is in development and will accommodate configuring three counters for each probe. Figure 7.9 shows screenshots of the user interface.

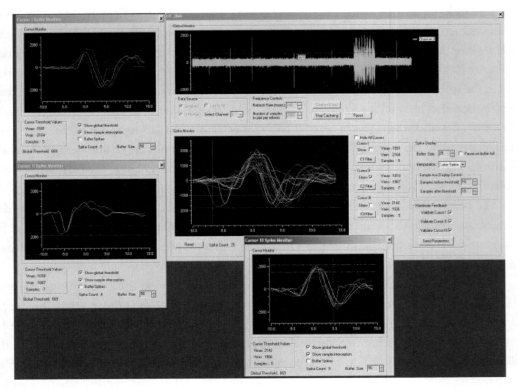

FIGURE 7.9: Spike detection parameter initialization application.

The Spike Rate Data Display application is used for demonstration purposes. In an actual laboratory environment, a Windows service reads the spike counts from the USB port and stores them in shared memory so any Windows application can process them.

7.3.1   In Vivo Testing and Results

Because invasive microwire recording technologies are targeting single-unit neuronal activity, we will use the action potential (spike) as our standard for evaluation of data collected from the neural probes [48]. The Pico system and custom amplifier have been previously tested extensively using a Bionic Neural Simulator to generate stereotypical physiological signals on the electronics workbench. In vivo tests used adult male 250-g Sprague–Dawley rats to benchmark the recording performance of the custom amplifier and Pico system. All procedures have been approved by the University of Florida IACUC Board. A 16-electrode (2 × 8) microwire array was implanted to a depth of 1.66 mm into the forelimb region of the primary motor cortex. The array was then grounded by using a 1/16-in.-diameter stainless steel screw implanted in the skull.

Figure 7.10 shows a 30-sec trace of neural recordings captured with the Pico system and gives a global perspective into the peak-to-peak amplitudes and occurrences of spiking neurons. The average of the data over the 30 sec is expected to be about zero. Instead, there is an offset of nearly 9 μV, which has been determined to be attributed to not precisely biasing the amplifiers.

In Figure 7.11, we zoom in on two individual action potentials where the range of the noise floor can take values of ±25 μV with the spikes reaching –75 μV.

CED's Spike 2 analysis software was used to extract neural data from the dataset. Figure 7.12 shows the average waveshape of two neurons firing on the probe from which we sampled.

It is difficult to calculate meaningful SNRs on neurophysiological data, because the spike signal is not continuous like those used in traditional communication's calculations [53]. Table 7.3 presents an attempt to derive meaningful SNR calculations from the average spike value. The RMS voltage of the noise floor was calculated by taking the square root of the average of the square of

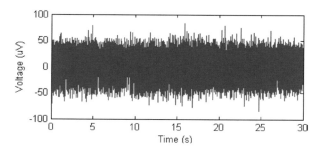

FIGURE 7.10: Thirty seconds of extracellular recordings from an anesthetized rat.

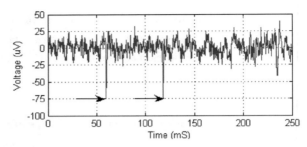

FIGURE 7.11: Two hundred fifty microseconds of extracellular recordings with two spikes.

samples in the noise floor. The samples used in the calculation were extracted by looking for periods of data where no spikes appeared.

To provide a baseline for quantifying the noise attributed by the system, the inputs to the probes were tied to the reference and the corresponding trace is given in Figure 7.13. These results suggest that the source of the dc offset is inherent in the system, which supports the need to carefully adjust the bias of the amplifiers. When these data are normalized, the RMS voltage of the noise using the same process described earlier yields 0.66 μV, which is much lower when compared to the 8.24 μV in the actual test.

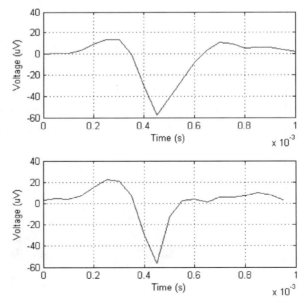

FIGURE 7.12: Average waveform spikes of two different neurons.

TABLE 7.3: SNR calculations		
	NEURON 1	**NEURON 2**
Noise V_{rms} (µV)	8.24	8.24
Neuron V_{pp} (µV)	71.13	79.33
SNR (dB)	18.72	19.67

For all of the neural recordings presented here, the system power consumption for the Pico Remote is quantified in Table 7.4. The Pico remote consumes around 130 mW, whereas bin counting, on a 3.7-V, 150 mA h LiON battery, would provide about 4.25 h of operation.

Now that the initial goal of bringing actual neural data from a live subject over the wireless link to a PC has been achieved, we can move forward in processing the data on the Pico remote by applying our spike counting algorithm. There are still additional tests that have to be completed, such as, comparing in a parallel test, raw signal data and spike counts with a commercial system, such as those from TDT Systems.

To ensure data integrity over the wireless link, we have created specifications for a robust communications protocol that utilizes packet acknowledgements and resends. Any data sent over the wireless network will be acknowledged by the receiving device. If the sending device does not receive an acknowledgement in a few milliseconds, the data are resent. A new version of the Nordic part allows for data resends in hardware reducing power and firmware complexity. Once the latest Nordic parts and the protocol are implemented, we expect data loss to be very close to zero.

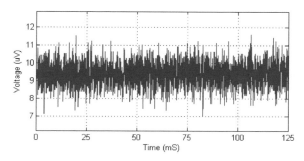

FIGURE 7.13: Noise introduced by the Pico system.

TABLE 7.4: Pico Remote power consumption		
SYSTEM	SPIKE SORTING (mW)	BIN COUNTING (mW)
Pico Remote	100	87
Pico Remote with TDT with headstage	143	130

Once the tests are completed and the design is stable, the system will be repackaged in a small rat backpack not larger than 1.5 × 3 in. in size and weighing a few ounces.

7.4   PORTABLE DSP DESIGNS: THE NEURAL SIGNAL PROCESSOR

Because the modeling is in the critical path of BMIs, an effort to create low-power flexible DSP boards was pursued in the scope of generation 1 implementations. The system described here serves as the digital portion of the overall BMI structure and is responsible for mapping the neural firings into action in the external world. To do this mapping, the system first acquires digitized neural data from the analog module. After acquisition and spike detection, the system then computes a linear or nonlinear model to map the neural data to a trajectory. Once a trajectory is determined, the system finally transmits commands wirelessly to an off-board robotic arm as described in mode 3. Mode 2 probably can be implemented without the use of a DSP. Mode 1 also requires a DSP for data compression. Recent results with a specially designed self-organizing map [54] or a vector quantizer tuned to action potentials [54] show that compression ratios of at least 100 are possible preserving spike shape.

Because the system needs to be carried by a subject (human, primate, or rat) without wires, there are portability and wireless constraints that will affect the design choices. First, the size and weight must be small enough to accommodate the various subject species. This involves the actual size of the components and the printed circuit board layout. Second, power consumption must be low to allow the use of lighter batteries and yield longer experimental trials. With power and size constraints placed on the system, the choice of processor and wireless transceiver becomes a concern. Because most of the prediction models require the use of floating point arithmetic, the processor needs to be floating point with enough speed to attain real-time model computations yet low power enough to appeal to the system constraints. In addition, the wireless transceiver must not only be small with a sufficient transmission bandwidth for current and near-future needs, it must also be low power. In the following subsections, we detail the NSP component choices that address these system requirements. Although these choices address the system requirements, the modular design allows for replacement of the individual components as new technologies are developed with better specifications (Figure 2).

7.4.1   Digital Signal Processor

The central component to any computational system is the processor. The processor has the ability to determine the speed, computational throughput, and power consumption of the entire system. Moreover, it dictates what additional support devices will be used for the overall system design. For the NSP, we chose the Texas Instruments (Dallas, TX) TMS320VC33 (C33) DSP. We chose this processor because it is suitable for on-board BMI modeling in terms of computational needs, power consumption, and versatility (Figure 7.14).

Specifically, the C33 meets high speed and floating-point requirements because it is a floating-point DSP capable of up to 200 mega-floating point operations per second (with over clocking). It achieves such high speeds by taking advantage of its dedicated floating point/integer multiplier. In addition, it works in tandem with the arithmetic logic unit, so that it has the ability to compute two

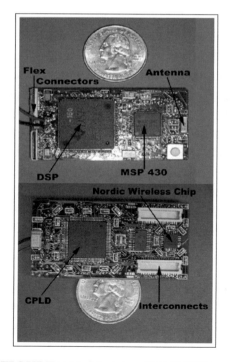

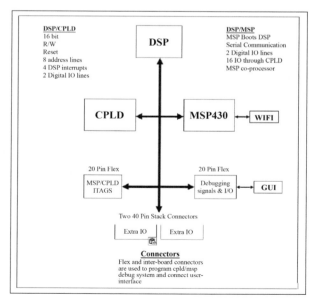

FIGURE 7.14: Modules of BMI DSP system.

mathematical operations in a single cycle. With support for both assembler and C/C++, code for the DSP can be optimized at low and high programming levels for additional speed support. Even at 200 MFLOPS, the C33 is also well suited to low power needs because it uses less than 200 mW. It achieves such power savings due in part to its 1.8-V core and other power-saving measures built into the processor. Using adjustable clock speeds, the DSP can exchange processing speed for lower power consumption. This may be desirable because some BMI experiments may require increased battery life in exchange for less computational speed (i.e., less neural data). Finally, the C33 is able to fulfill other expandability requirements of the system. First, it supports a large address space by providing a 24-bit address bus to read/write 16 million different locations. This address space is used to map the different hardware components of the system and future hardware interfaces. It includes a 32-bit parallel data bus and serial port for multiple communication methods and hardware multiplexing. This processor can also provide expansion because it has four hardware interrupts, two 32-bit timers, and a direct memory access controller that can be used for future requirements or hardware interfaces.

Wireless Communication.   As with the PICO system, the wireless connection is the second most important hardware module in the NSP. This module, similar to the processor, has the ability to define the size and power consumption of the full system. Early on, we determined that 802.11B was the most convenient protocol for our group and collaborators [18]; unfortunately, it required a PCMCIA card that was very large and consumed the majority of the power from the system. Because the digital module does not need a lot of bandwidth to transmit 100-msec trajectory co-ordinates (20 bytes), and the possible connection to the analog frontend if wireless uses at most 500 kbits/sec, we sacrifice the bandwidth of the PCMCIA card and 802.11Bs universal protocol for a tiny low power 2.4-GHz wireless solution. This wireless chip (also used in the PICO), Nordic nRF2401, is a single chip transceiver capable of 1 Mbp/sec at 10.5 mA (peak). It also has the ability to reduce its power further to 0.8 mA if we use ShockBurst mode or any variant in between. Compared to the 100- to 300-mA PCMCIA card (depending on throughput), this is a significant savings in power. In addition, the package size of the Nordic chip is only 5 × 5 mm compared to the 4.5 × 2.25-in. PCMCIA card used in the previous design. Although there is an additional antenna, it is only a 6.5 × 2.2 mm surface mount SMD, still making the Nordic chip a significant savings in size and power.

MSP Coprocessor/Boot-Loader.   The NSP contains a coprocessor to alleviate the computational strain on the DSP. In turn, this coprocessor supports the current peripheral functions of the digital module and any future functions. Consequently, this allows the DSP to solely concentrate on the algorithms for neural-to-trajectory mappings. We chose the MSP430F1611x to serve as the coprocessor because it is a very low power yet highly functional microcontroller. This 16-bit RISC microcontroller consumes 1 µA of power at 1 MHz and linearly increases power to its maximum at 8 MHz. In addition, there are other power modes that allow it to use infinitesimally small amounts of power with voltage ranges from 1.8 to 3.6 V. For the NSP, we use 3.3 V to supply the microcontroller. Furthermore, the microcontroller has 48 KB of flash RAM that can store data when power is off and 10 KB when the power is on. Combined with a 12A/D-D/A and multiple asynchronous/synchronous serial ports and SPI/I2C ports, this microcontroller is very well endowed for such a small package (10 × 10 mm).

Handling the wireless protocol is one of the main functions this coprocessor is in charge of. It provides the DSP communication transparency when sending and receiving wireless data. Underneath this transparency the multifunctional component (MSP) is controlling the flow and control of the wireless chip. We also use the MSP as a boot loader for the DSP by using its serial port for booting. To boot the DSP serially, we store the DSP program in the flash because the MSP has 48 KB of flash. Essentially, by using the MSP to boot the DSP, we trade one single functioned component for an MSP that is lower power. In terms of expandability, the MSP is fast enough (8 million instructions per second) to handle preprocessing of the neural data for the prediction models running on the DSP. Through a serial port or a digital IO (connected to the flex and interconnects), we can also expand the system to include USB communications for the uploading or downloading of data in real time (from multiple paths MSP/DSP/analog module). We can also use these ports to create a test bed for debugging the system or testing future analog modules. Finally, similar to the wireless chips, we have the ability to scale the NSP system and add multiple coprocessors because they are very low power and small, yet highly functional.

Reconfigurability.   1) Complex Programmable Logic Device (CPLD). By using an Altera C3000, the NSP is able to expand and accommodate future needs. There are options for using on-board clocks as well as off-board and even clocks from the analog module. This makes for a very flexible system for future analog modules that require or provide clock signals. Communication is also reconfigurable because it is critical between the components on-board and modules off board. This includes the serial ports from the DSP and MSP, and the parallel data bus from both processors. Through the CPLD we have the ability to use a USB connection and other serial ports. These in turn can be redirected to the DSP or the MSP. The NSP even has the ability to redirect the wireless connection directly to the external IO for debugging and support. Essentially, the CPLD is very useful for debugging the system and add future features that may become required. It has access to all the connectors and components. Because it is reprogrammable through very high speed integrated circuit hardware (VHDL), it makes for a very versatile hardware component.

2) Connector Interface. To complement the CPLD and allow for functionality of unknown future needs it is necessary to provide as much connectivity between the internal components to the

external. Coupled with the desire to be portable we need connecters that have many pins and yet small enough to not enlarge the system. In addition, we need to be able to provide support functionality such as debugging and test beds.

The first connection type must be able to mate with the analog modules and provide a suitable amount of connections, and yet be strong and compact enough to support another attached printed circuit board. For the NSP, we use two connectors that allow for 80 different connections between the boards. With this amount of connections, we can not only support current needs from an analog module [duke], we can also support other daughter boards that may be designed for testing or future boards.

The second connection type is more inline with debugging and programming the DSP, CPLD, and MSP. In addition, to retain the ability to communicate to systems that may be at great distances we chose to design a flex connecter interface. On these connectors we provide access to the CPLD jtag, MSP serial ports, wireless serial port, and other general CPLD IO for multiplexing communication between the components of the NSP. Finally, the connector layout is such that it allows for stackable modules, providing flexibility for all the three modes of operations described above.

3) Power Subsystem. The two different voltage requirements for the NSP are 1.8 and 3.3 V. These voltages are not only used by the DSP, they are also used for the other hardware modules in the system. In trying to supply the NSP components, the voltages can be supplied from multiple pathways to the NSP. This includes an on-board option and off-board. The main option for power supplying is from on-board. This includes a Texas Instruments TPS70351 Dual-Output LDO Voltage Regulator that provides both 1.8 and 3.3 V voltages on a single chip and only requires 5 V to operate. This chip also provides the power-up sequence required by the DSP once it is initialized. In addition, requiring a single supply is very desirable for a portable system because only one 5-V battery supply is necessary. With the off-board option, we can power the regulator through the flex connectors, interconnect connector or a simple header; this way, a battery or external supply can power the regulator. In addition, we can have the analog module or debugging system provide the individual power requirements. This allows for the isolation of certain components for being powered. With such variability, future hardware can take advantage of the different powering options.

System Software.   There are six layers of software in the NSP system environment: 1) PC software, 2) DSP operating system (OS), 3) DSP algorithms, 4)VHDL code 5) MSP430 code, and 6) client/server code. We wrote a PC console program to interface the DSP through the USB. The console program calls functions within the DSP OS to initiate and control the USB communication functions. The DSP OS is also responsible for reading/writing memory locations and various program control functions. This is the main pathway for debugging the system. In tandem with the OS layer of the DSP, there is low-level MSP code for initializing and controlling the wireless controller. This code must interact with the DSP OS and any client code or algorithms that are running simultaneously within the NSP. Once the prediction model completes an epoch or computation cycle, the program must interrupt the MSP and transfer any required data. This process also involves creating the correct packets for transmission to the appropriate server (off-board).

The final layer of code resides in the on-board CPLD. This hardware-based VHDL code is responsible for correctly shuttling data between all of the hardware modules. It achieves this processing through a series of interrupts and control lines that are provided by the individual hardware components. Depending on the functionality of the system, this code is responsible for all the internal and external communication of the NSP system.

We use a high-speed USB connection to transmit 104 binned neural channels and 3D trajectory data to compute an optimal linear combiner using 10-tap FIRs per channel trained with the normalized LMS algorithm for a total of 3120 parameters [55]. The average amount of time for a single prediction when computed over 90 predicted outputs is 211 μsec. This test achieves the same timing results on the computation of predicted output because this was the same DSP used in our previous systems [56, 57]. On a 600-MHz PIII laptop running Matlab 6, the average prediction takes 23 msec. The factor of improvement or speed gain is about 100× for the DSP over the laptop running Matlab. The LMS output results collected at the receiving computer were directly compared to Matlab computed outputs and they agree within seven decimal places. The NSP uses approximately 90 to 120 mA, which equates to 450–600 mW. Overall, this is much lower than the 4 and 1.75 W previously attained by other acquisition hardware [56, 57]. In addition, the board size is 2.5 × 1 in. board size and weighs 9 g.

7.5   FLORIDA WIRELESS IMPLANTABLE RECORDING ELECTRODES

The second-generation systems will have to deal with an immensely more restrictive set of requirements because they will be subcutaneous. We envision a modular design with tens to hundreds of channels. The present requirements that we are working with is an implantable rechargeable device called the Florida wireless implantable recording electrodes (FWIRE) that will measure 1.5 × 1 × 0.5 cm, weigh 2 g, and can collect data from 16 microelectrodes for 100 hours, amplify and send them wirelessly within 500 Kbits/sec to a remote system within 1 m of the subject. A conceptual building block is presented in Figure 7.15.

As illustrated in Figure 7.15, the FWIRE microsystem consists of a flexible substrate that serves as the platform for the signal processing integrated and fine integrated circuit and wireless telemetry radio frequency integrated circuit chips, transmit antenna, receive and power coil, and microwire electrode array. A low-profile rechargeable battery is located below the flexible substrate and is used to power the implant electronics during recording sessions. The external coil wraps

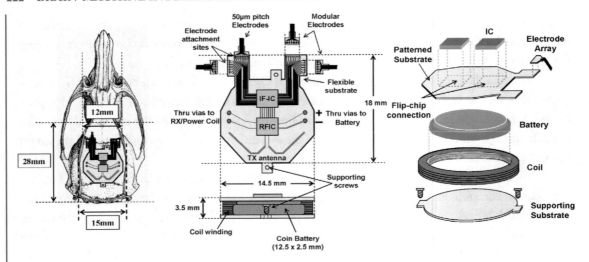

FIGURE 7.15: Component diagram and assembly of the Florida wireless implantable recording electrode system.

around the battery for a very compact package assembly and enables wireless battery recharging. The CMOS ICs are flipchip-bonded onto the flexible substrate, which is encased in medical grade silicone for isolation from fluids. Mechanical stability of the patterned flexible substrate is provided by the underlying battery/coil, supporting substrate, and screws, which attaches the platform to the skull of the behaving animal and provides the required ground reference for the on-chip electronics. Multiple electrode attachment sites onto the flexible substrate using micromachined flexible polyimide ribbon cable provide additional flexibility for experimentation. The metal traces and corresponding bond sites can be made to any size specification and spacing distance via photolithography. The initial prototypes of the electrode array shown in Section 7.1 consists of a row of eight (30 μm diameter with 50 μm pitch) gold-plated nickel electrodes with parylene-C insulated shanks that extend 4 mm from the edge of the flexible polyimide cable [58]. In vivo studies produced a high neuronal yield with SNR ranging from 10 to 27 dB. The resulting FWIRE platform is a highly modular architecture that takes into consideration the mechanical and physical constraints of the implant site, and facilitates independent development and testing of its individual components.

7.5.1 Neural Signal Processing and Representation

To achieve these objectives, very fundamental aspects of both the electronics and signal processing must be modified. We believe that a hybrid analog–digital signal processing approach is necessary, and started some preliminary steps in that direction. The first is to rethink Nyquist sampling, which is lopsided towards a complex frontend (accurate sample-and-hold that uses large chip areas and consumes a lot of power) versus a very easy backend (lowpass filtering). Here we require exactly the opposite, that is, a sampling scheme that is implemented in a simple frontend circuit eventually with a more difficult reconstruction in the backend. Our group at Florida has recently proposed the integrate-and-fire (IF) representation, which accomplishes these goals. The signal is integrated up to a threshold and a pulse is fired when the voltage crosses a threshold. Hence, the time between the pulses is proportional to the integral of the voltage, providing an aperiodic sampling scheme. Provided that there are sufficient pulses on an interval, one can show that this scheme obeys the definition of sampling (a one to one mapping with a unique inverse). What is interesting is that the IF representation is very easy to build in hardware and is immune to noise in the chip on during transmission because of the pulse nature of the representation. As long as delays are constants, they do not affect the accuracy of the representation. The quantization noise is associated with how precise is the pulse timing, so data rates will not increase with precision as in conventional A/Ds. Moreover, this representation naturally improves the fidelity of the high-amplitude portions of the signals (where there are more spikes), so in many applications (as in spike trains) considerably lower data rates are necessary to represent the signals when compared with Nyquist samplers. The power–size–bandwidth specification of this sampler seems better than conventional Nyquist samplers.

The availability of the IF representation also opens the door to blend analog and signal processing in a much more interesting way. Figure 7.16 shows our present thinking on three approaches to decrease data rates through processing at the frontend.

Strategy 1 is sends out the data directly after the IF representation. The data rates depend on the input structure and the threshold. By selecting higher thresholds the data rates decrease, but aliasing starts cropping up in the low amplitude portions of the signal.

Strategy 2 does signal compression in the pulse domain before sending the data out. The idea is to use the time between the pulses and their polarity to recognize the events of importance that in our case will be neural spikes. The objective is to decrease the data rates and only send out events that are of interest.

The third strategy is to ultimately do the processing on the chip by creating signal processing algorithms that work with point processes. The output will be directly connected to an external device for brain control.

Integrate and Fire Representation.   The voltage output of the amplifier is first converted into current and, by integrating this current, the amplitude information is encoded into an asynchronous digital pulse train. This IF module efficiently encodes the analog information as an asynchronous train of pulses [59]. The principle is to encode amplitude in the time difference between events. The advantage is that a single event represents a high-resolution amplitude measurement, with

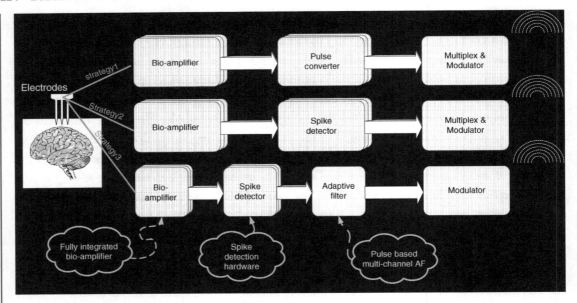

FIGURE 7.16: Three strategies for contending with bandwidth limitations of wireless implantable neural recording systems.

substantial bandwidth and power savings. These pulses have better noise immunity than conventional analog signals in transmission and also eliminates the need for a traditional ADC. The latest version using a delta–sigma modulation style to produce timing events resulted in 52 μW of power dissipation with 8-bit resolution [60]. Given a band-limited continuous signal $y(t)$, we create an alternate representation

$$\hat{y}(t) = h(t) * \sum_k w_k \delta(t - t_k) = \sum_k w_k h(t - t_k) \qquad (7.1)$$

where $|y(t) - \hat{y}(t)| <_\varepsilon$ for all time, w_k are appropriate weights, and $h(t)$ is the sinc function. Sampling times t_i are given by: $f(t_k) = k_\theta$, where

$$\int_{t_k}^{t_{k+1}} y(t)\mathrm{d}t = \theta \text{ and } f(t) = \int_{t_0}^{t} y(\tau)\mathrm{d}\tau \qquad (7.2)$$

Previously, simple linear low-pass filters were used as crude reconstruction filters from IF outputs, computing something similar to rate coding from the spike train, and the performance was poor. Pulse times are determined by the local area under the signal input $x(t)$, and they can be used for perfect reconstruction. Let us assume that $x(t)$ is band-limited to Ω_s and the maximum interspike interval $T_{\max}$ satisfies $T_{\max} < \pi/\Omega_s$. From mathematical frame theory, it is known that any band-limited signal can be expressed as a lowpass filtered version of an appropriately weighted sum of delayed impulse functions, which is expressed in equation (1). So the signal recovery problem is how to calculate the appropriate weights. We can define the timing of pulses: $T = [t_1, t_2, \ldots, t_k, t_{k+1}, \ldots]$. Let $(t) = v_{in}A_{in} - V_{ref\text{-}supply} + V_{mid}$. Substituting equation (7.2) into equation (7.1), we obtain

$$\theta = \int_{t_i}^{t_{i+1}} y(t)\,\mathrm{d}t = \int_{t_i}^{t_{i+1}} \sum_j w_j h(t - t_j)\,\mathrm{d}t = \sum_j w_j \int_{t_i}^{t_{i+1}} w_j h(t - t_j)\,\mathrm{d}t \tag{7.3}$$

$$c_{ij} = \int_{t_i}^{t_{i+1}} h(t - t_j)\,\mathrm{d}t \tag{7.4}$$

which can be rewritten in matrix notation as $CW = \theta$. Unfortunately, C is usually ill-conditioned, necessitating the use of a SVD-based pseudoinverse to calculate the weight vector **W**. We have extended the reconstruction algorithm to work for a biphasic pulse train with positive and negative pulse indicating when a positive or negative threshold is surpassed [59].

7.5.2   Wireless

A wireless interface for the FWIRE microsystem is currently being developed to facilitate in vivo neural recording in behaving animals. As shown in Figure 7.17, the system includes a power recovery, regulation, and battery management module; an ASK clock and data recovery (CDR) circuit to download external system commands; a transmitter and signal processing circuits to upload neural data from the various recording channels; and a small controller with register bank to oversee system functionality, decode/encode data packets, and store on-chip settings. All components are fully integrated with the exception of the power/downlink coil, the uplink antenna, and the battery.

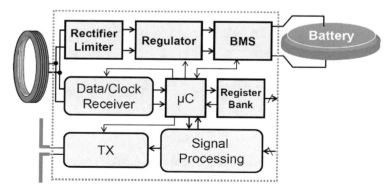

FIGURE 7.17: Wireless interface system block diagram.

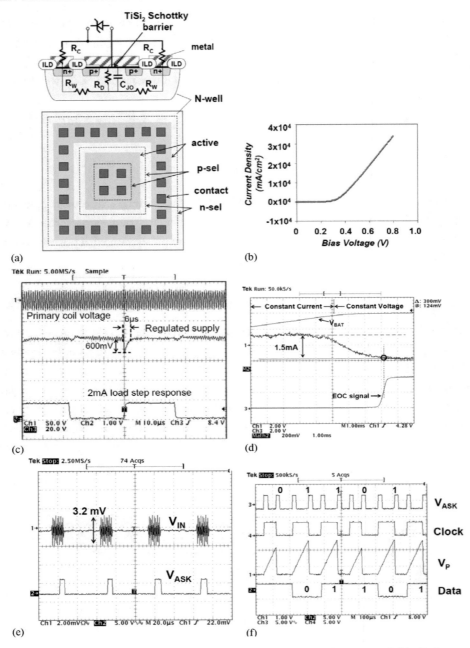

FIGURE 7.18: (a) Schottky contact barrier diode in standard CMOS; measured (b) diode current density versus bias voltage; (c) regulator transient response; (d) battery control loop charging profiles; (e) ASK detector; (f) CDR waveforms.

7.5.2.1   Wireless Power and Data Interface

A 4-MHz LC tank with on-chip resonant capacitor, full-wave rectifier, storage capacitor, low drop-out voltage regulator, and RF limiting circuit provide a stable supply and overvoltage protection to the implant electronics. Schottky contact barrier diodes are used in the bridge rectifier circuit to improve the frequency response and lower the turn-on voltage. As shown in Figure 7.18a, Ti-Si$_2$ Schottky barrier diodes are fabricated in standard 0.6-μm CMOS process by selectively blocking the n+/p+ implants in desired diffusion areas [61]. The measured I–J characteristic of the Schottky barrier devices is shown in Figure 7.18b. The devices have an ideality factor of ~1.16 and barrier height of ~0.5 eV, which was computed using the Richardson–Dushman equation for the thermionic current [62]. The measured turn-on voltage is ~300 mV, which is lower than standard pn junction based implementations [63]. The diodes are fabricated with small schottky contact area as this yields a higher cutoff frequency [64], and multiple cells are placed in parallel to improve the current handling capability. Figure 7.18c shows the transient regulator response when an externally generated 0- to 2-mA load step is applied as the link is powered by the primary coil voltage. The measured response is within 15% (or 600 mV/4.1 V) of the target 4.1-V supply. The regulator exhibits a load regulation of 2 mV/mA (or 240 ppm/1 mA), a line regulation 2 mV/V, and a low dropout voltage of 50 mV. At 4 MHz, the worst-case measured peak-to-peak ripple voltage at the output is within 100 mV. A battery management circuit has been implemented to charge and monitor the external Li ion cell battery [61]. Charging profile experiments for the battery control loop (Figure 7.18d) shows that during the constant-current phase, the circuit delivers 1.5 mA, resulting in a linear increase in battery voltage. The end-of-charge (EOC) during the constant-voltage phase is detected once the battery current reaches 5% of the nominal constant charging current of 1.5 mA, triggering the EOC signal. An on-chip CDR circuit is used to download external system commands. Synchronization of clock and data is achieved via a modulation scheme based on amplitude shift keying and pulse position modulation facilitates CDR using an ASK demodulator and a charge-pump with latched comparator. The CDR is operational for an input data range of 4–18 kb/sec, and exhibits a sensitivity of 3.2 mV p-p at 1 MHz.

7.6   SUMMARY

The breadth of challenges that stand between concept generation and realization of the next generation of BMIs calls for tandem and synergistic development of the many facets of neurotechnology. Here, we have provided several examples of how the goal of system development (as opposed to component development) of a fully integrated BMI implant is leading to beyond-state-of-the art technologies that might not be realized readily in a piecewise fashion. To develop the necessary signal processing, neural interfaces, or electronics independently, as has been largely conducted in the past, may not yield optimal BMI systems that are capable of chronically interfacing with large neural assemblies and delivering meaningful function to the patient. The complete specifications of future BMI systems will be driven by the technical challenges encountered in a particular specialization of BMI neurotechnology and transferred bidirectionally to the others. For example, the choice of the appropriate physiological scale to derive the most robust control hinges upon the electrode design and polymer science to sense signals throughout the life of the patient. However, without feedback on the performance of neural decoding, one may have the nonoptimal but biocompatible electrodes that do not provide the most relevant information. It is clear that the next generation of BMI technologies cannot be built solely from existing engineering principles. So the question then becomes, "What is the best strategy to deal with technology bottlenecks?" Fortunately, neural engineering approaches can benefit from the intersection of top–down signal processing techniques with bottom–up theories of neurophysiology. By reverse engineering the neural interface problem using modified versions of standard signal processing techniques, one can overcome the bottlenecks.

REFERENCES

1. Xu, S., et al., *A multi-channel telemetry system for brain microstimulation in freely roaming animals.* Journal of Neuroscience Methods, 2004. **133**: pp. 57–63. doi:10.1016/j.jneumeth.2003.09.012

2. Beechey, P., and D.W. Lincoln, *A miniature FM transmitter for radio-telemetry of unit activity.* Journal of Physiology (London), 1969. **203**(1): p. 5.

3. Edge, G.M., G. Horn, and G. Stechler, *Telemetry of single unit activity from a remotely controlled microelectrode.* Journal of Physiology (London), 1969. **204**(1): p. 2.

4. Eichenbaum, H., et al., *Compact miniature microelectrode-telemetry system.* Physiology & Behavior, 1977. **18**(6): pp. 1175–1178. doi:10.1016/0031-9384(77)90026-9

5. Grohrock, P., U. Hausler, and U. Jurgens, *Dual-channel telemetry system for recording vocalization-correlated neuronal activity in freely moving squirrel monkeys.* Journal of Neuroscience Methods, 1997. **76**(1): pp. 7–13. doi:10.1016/S0165-0270(97)00068-X

6. Strumwasser, F., *Long-term recording from single neurons in brain of unrestrained mammals.* Science, 1958. **127**(3296): pp. 469–470.

7. Warner, H., et al., *A remote control brain telestimulator with solar cell power supply.* IEEE Transactions on Biomedical Engineering, 1968. **BM15**(2): p. 94.

8. Akin, T., et al., *A modular micromachined high-density connector system for biomedical applications.* IEEE Transactions on Biomedical Engineering, 1999. **46**(4): pp. 471–480. doi:10.1109/10.752944

9. Bai, Q., and K.D. Wise, *Single-unit neural recording with active microelectrode arrays*. IEEE Transactions on Biomedical Engineering, 2001. **48**(8): pp. 911–920.

10. Berger, T.W., et al., *Brain-implantable biomimetic electronics as the next era in neural prosthetics*. Proceedings of the IEEE, 2001. **89**(7): pp. 993–1012. doi:10.1109/5.939806

11. Blum, N.A., et al., *Multisite microprobes for neural recordings*. IEEE Transactions on Biomedical Engineering, 1991. **38**: pp. 68–74. doi:10.1109/10.68211

12. Campbell, P.K., et al., *A silicon-based, 3-dimensional neural interface—manufacturing processes for an intracortical electrode array*. IEEE Transactions on Biomedical Engineering, 1991. **38**(8): pp. 758–768.

13. Chang, J.C., G.J. Brewer, and B.C. Wheeler, *Modulation of neural network activity by patterning*. Biosensors and Bioelectronics, 2001. 16(7–8): pp. 527–533. doi:10.1016/S0956-5663(01)00166-X

14. Jones, K.E., and R.A. Normann, *An advanced demultiplexing system for physiological stimulation*. IEEE Transactions on Biomedical Engineering, 1997. **44**(12): pp. 1210–1220. doi:10.1109/10.649992

15. Kennedy, P.R., *The cone electrode—a long-term electrode that records from neurites grown onto its recording surface*. Journal of Neuroscience Methods, 1989. 29(3): pp. 181–193. doi:10.1016/0165-0270(89)90142-8

16. Kruger, J., *A 12-fold microelectrode for recording from vertically aligned cortical-neurons*. Journal of Neuroscience Methods, 1982. **6**(4): pp. 347–350. doi:10.1016/0165-0270(82)90035-8

17. Moxon, K.A., et al., *Ceramic-based multisite electrode arrays for chronic single-neuron recording*. IEEE Transactions on Biomedical Engineering, 2004. **51**(4): pp. 647–656. doi:10.1109/TBME.2003.821037

18. Obeid, I., M.A.L. Nicolelis, and P.D. Wolf, *A low power multichannel analog front end for portable neural signal recordings*. Journal of Neuroscience Methods, 2004. **133**(1–2): pp. 27–32. doi:10.1016/j.jneumeth.2003.09.024

19. Palmer, C.I., *Long-term recordings in the cat motor cortex unit-activity and field potentials from sensory and brain-stem stimulation as a means of identifying electrode position*. Journal of Neuroscience Methods, 1990. **31**(2): pp. 163–181. doi:10.1016/0165-0270(90)90161-8

20. Salcman, M., *Trauma after electrode implantation*. Archives of Neurology, 1976. **33**(3): pp. 215–215.

21. Tsai, M.L., and C.T. Yen, *A simple method for fabricating horizontal and vertical microwire arrays*. Journal of Neuroscience Methods, 2003. **131**(1–2): pp. 107–110. doi:10.1016/S0165-0270(03)00235-8

22. Wise, K.D., et al., *Wireless implantable microsystems: High-density electronic interfaces to the nervous system*. Proceedings of the IEEE, 2004. **92**(1): pp. 76–97. doi:10.1109/JPROC.2003.820544

23. Buzsaki, G., *Multisite recording of brain field potentials and unit activity in freely moving rats.* Journal of Neuroscience Methods, 1989. **28**: pp. 209–217. doi:10.1016/0165-0270(89)90038-1

24. Farshchi, S., et al., *A TinyOS-enabled MICA2-based wireless neural interface.* IEEE Transactions on Biomedical Engineering, 2006. **57**(3): pp. 1416–1424. doi:10.1109/TBME.2006.873760

25. Mavoori, J., et al., *An autonomous implantable computer for neural recording and stimulation in unrestricted primates.* Journal of Neuroscience Methods, 2005. **148**(1): pp. 71–77.

26. Abeles, M., Corticonics: Neural Circuits of the Cerebral Cortex. 1991, New York: Cambridge University Press.

27. Wessberg, J., et al., *Real-time prediction of hand trajectory by ensembles of cortical neurons in primates.* Nature, 2000. **408**(6810): pp. 361–365.

28. Moxon, K.A., S.C. Leiser, and G.A. Gerhardt, *Ceramic-based multisite electrode arrays for chronic single-neuron recording.* IEEE Transactions on Biomedical Engineering, 2004. **51**(4): pp. 647–656. doi:10.1109/TBME.2003.821037

29. Chapin, J.K., and K.A. Moxon, eds. *Neural prostheses for restoration of sensory and motor function.* Methods and New Frontiers in Neuroscience. 2001, Boca Raton, FL: CRC Press.

30. Kipke, D.R., R.J. Vetter, and J.C. Williams, *Silicon-substrate intracortical microelectrode arrays for long-term recording of neuronal spike activity in cerebral cortex.* IEEE Transactions on Rehabilitation Engineering, 2003. **11**(2): pp. 151–155. doi:10.1109/TNSRE.2003.814443

31. Cham, J.G., et al., *Semi-chronic motorized microdrive and control algorithm for autonomously isolating and maintaining optimal extracellular action potentials.* Journal of Neurophysiology, 2005. **93**(1): pp. 570–579. doi:10.1152/jn.00369.2004

32. Rousche, P.J., and R.A. Normann, *A method for pneumatically inserting an array of penetrating electrodes into cortical tissue.* Annals of Biomedical Engineering, 1992. **20**: pp. 413–422. doi:10.1007/BF02368133

33. Jaeger, D., S. Gilman, and J.W. Aldridge, *A multiwire microelectrode for single unit recording in deep brain structures.* Journal of Neuroscience Methods, 1990. **32**(2): pp. 143–148. doi:10.1016/0165-0270(90)90170-K

34. Jellema, T., and J.A.W. Weijnen, *A slim needle-shaped multiwire microelectrode for intracerebral recording.* Journal of Neuroscience Methods, 1991. **40**(2–3): pp. 203–209. doi:10.1016/0165-0270(91)90069-C

35. Williams, J.C., R.L. Rennaker, and D.R. Kipke, *Long-term neural recording characteristics of wire microelectrode arrays implanted in cerebral cortex.* Brain Research Protocols, 1999. **4**(3): p. 303. doi:10.1016/S1385-299X(99)00034-3

36. Ji, J., and K.D. Wise, *An implantable CMOS circuit interface for multiplexed microelectrode recording arrays.* IEEE Journal of Solid-State Circuits, 1992. **27**(3): pp. 433–443. doi:10.1109/4.12156

37. Najafi, K., and K.D. Wise, *An implantable multielectrode array with on-chip signal processing.* IEEE Journal of Solid State Circuits, 1986. **21**(6): pp. 1035–1044.

38. Maynard, E.M., C.T. Nordhausen, and R.A. Normann, *The Utah Intracortical Electrode Array: A recording structure for potential brain–computer interfaces.* Electroencephalography and Clinical Neurophysiology, 1997. **102**(3): pp. 228–239. doi:10.1016/S0013-4694(96)95176-0

39. Nordhausen, C.T., P.J. Rousche, and R.A. Normann, *Optimizing recording capabilities of the Utah-Intracortical-Electrode-Array.* Brain Research, 1994. **637**(1–2): pp. 27–36. doi:10.1016/0006-8993(94)91213-0

40. Rousche, P.J., and R.A. Normann, *Chronic recording capability of the Utah Intracortical Electrode Array in cat sensory cortex.* Journal of Neuroscience Methods, 1998. **82**(1): p. 1. doi:10.1016/S0165-0270(98)00031-4

41. Fee, M.S., and A. Leonardo, Miniature motorized microdrive and commutator system for chronic neural recording in small animals. Journal of Neuroscience Methods, 2001. **112**(2): p. 83.

42. Subbaroyan, J., D.C. Martin, and D.R. Kipke, *A finite-element model of the mechanical effects of implantable microelectrodes in the cerebral cortex.* Journal of Neural Engineering, 2005. **2**: pp. 103–113. doi:10.1088/1741-2560/2/4/006

43. Rousche, P.J., et al., *'Flexible' polyimide-based intracortical electrode arrays with bioactive capability.* IEEE Transactions on Biomedical Engineering, 2001. **48**(3): pp. 361–371. doi:10.1109/10.914800

44. Vetter, R.J., et al., *Chronic neural recording using silicon-substrate microelectrode arrays implanted in cerebral cortex.* IEEE Transactions on Biomedical Engineering, 2004. **51**(6): pp. 896–904. doi:10.1109/TBME.2004.826680

45. Paxinos, G., The Rat Brain in Stereotaxic Coordinates. 1997, Sydney: Academic Press. doi:10.1016/0165-0270(80)90021-7

46. Lewicki, M.S., *A review of methods for spike sorting: The detection and classification of neural action potentials.* Network Computation in Neural Systems, 1998. **9**(4). pp. R53–R78. doi:10.1088/0954-898X/9/4/001

47. Wood, F., et al., *On the variability of manual spike sorting.* IEEE Transactions on Biomedical Engineering, 2004. **51**(6): pp. 912–918. doi:10.1109/TBME.2004.826677

48. Nicolelis, M.A.L., Methods for Neural Ensemble Recordings. 1999, Boca Raton, FL: CRC Press.

49. Harrison, R.R., and C. Charles, *A low-power low-noise CMOS amplifier for neural recording applications.* IEEE Journal of Solid-State Circuits, 2003. **38**: pp. 958–965.

50. Chen, D., An Ultra-Low Power Neural Recording System Using Pulse Representations. 2006, Gainesville, FL: Department of Electrical and Computer Engineering, University of Florida.

51. Li, Y., An Integrated Multichannel Neural Recording System With Spike Outputs. 2006, Gainesville, FL: Department of Electrical and Computer Engineering, University of Florida.

52. Cieslewski, G., et al. *Neural signal sampling via the low power wireless Pico system*, in IEEE International Conference of the Engineering in Medicine and Biology Society. 2006. New York.

53. Nenadic, Z., and J. Burdick, *Spike detection using the continuous wavelet transform.* IEEE Transactions on Biomedical Engineering, 2005. **52**(1): pp. 74–87. doi:10.1109/TBME.2004.839800

54. Cho, J., et al., *Self-organizing maps with dynamic learning for signal reconstruction.* Neural Networks, 2007. **20**(2): pp. 274–284. doi:10.1016/j.neunet.2006.12.002

55. Sanchez, J.C., et al. *Interpreting neural activity through linear and nonlinear models for brain machine interfaces*, in International Conference of Engineering in Medicine and Biology Society. 2003. Cancun, Mexico. doi:10.1109/IEMBS.2003.1280168

56. Shalom, D., et al. *A reconfigurable neural signal processor (NSP) for brain machine interfaces*, in Engineering in Medicine and Biology Society, 2006. EMBS '06. 28th Annual International Conference of the IEEE. 2006.

57. Darmanjian, S., et al. *A portable wireless DSP system for a brain machine interface*, in Neural Engineering, 2005. Proceedings of the 2nd International IEEE EMBS Conference. 2005.

58. Patrick, E., et al. *design and fabrication of a flexible substrate microelectrode array for brain machine interfaces*, in IEEE International Conference of the Engineering in Medicine and Biology Society. 2006. New York.

59. Chen, D., et al. *Asynchronous biphasic pulse signal coding and its CMOS realization*, in Proceedings of IEEE International Symposium on Circuits and Systems (ISCAS). 2006. Kos, Greece.

60. Wei, D., V. Garg, and J.G. Harris. *An asynchronous delta–sigma converter*, in Proceedings of IEEE International Symposium on Circuits and Systems (ISCAS). 2006. Kos, Greece.

61. Li, P., J. Principe, and R. Bashirullah, *A wireless power interface for rechargeable battery operated neural recording implants*, in IEEE Engineering in Medicine and Biology Conference. 2006. New York.

62. Shur, M., Physics of Semiconductor Devices. 1990, New York: Prentice-Hall.

63. Ghovanloo, M., and K. Najafi, *A modular 32-site wireless neural stimulation microsystem.* IEEE Journal on Solid-State Circuits, 2004. **39**(12): pp. 2457–2466.

64. Sankaran, S., and K.K. O, *Schottky barrier diodes for millimeter wave detection in a foundry CMOS process.* IEEE Electron Device Letters, 2005. **26**(7): pp. 492–494. doi:10.1109/LED.2005.851127

• • • •

Author Biography

Justin C. Sanchez Dr. Sanchez's research interests are in Neural Engineering and neural assistive technologies. Topics include the analysis of neural ensemble recordings, adaptive signal processing, Brain-Machine Interfaces, motor system electrophysiology, treatment of movement disorders, and the neurophysiology of epilepsy. He is an Assistant Professor of Pediatrics, Neuroscience, and Biomedical Engineering at the University of Florida College of Medicine, Engineering, and McKnight Brain Institute in Gainesville, Florida. He received his Ph.D. (2004) and M.E. degrees in Biomedical Engineering and B.S. degree in Engineering Science (Highest Honors—2000) with a minor in Biomechanics from the University of Florida. The goal of his research is to develop state-of-the-art novel medical treatments by operating at the interface between basic neural engineering research and clinical care. This direction of research is motivated by the potential of direct neural interfaces for delivering therapy and restoring functionality to disabled individuals using engineering principles. In 2005, he won two prestigious awards for his work including Excellence in Neuroengineering and more recently an American Epilepsy Society Young Investigator Award. In 2006 he founded the Gainesville Engineering in Medicine and Biology/Communications Joint Societies Chapter and serves as the IEEE Gainesville Section Director for membership development. His neural engineering electrophysiology laboratory is currently developing direct neural interfaces for use in the research and clinical settings and has published over 35 peer review papers and holds 3 patents in neuroprosthetic design. He is the founding member of the Neuroprosthetics Group (NRG) at the University of Florida (http://nrg.mbi.ufl.edu).

Dr. José C. Príncipe is Distinguished Professor of Electrical and Biomedical Engineering at the University of Florida since 2002. He joined the University of Florida in 1987, after an eight year appointment as Professor at the University of Aveiro, in Portugal. Dr. Principe holds degrees in electrical engineering from the University of Porto (Bachelor), Portugal, University of Florida (Master and Ph.D.), USA and a Laurea Honoris Causa degree from the Universita Mediterranea in Reggio Calabria, Italy. Dr. Principe's interests lie in nonlinear non-Gaussian optimal signal processing and modeling and in biomedical engineering. He created in 1991 the Computational NeuroEngineering Laboratory to synergistically focus the research in biological information processing models. He recently received the Gabor Award from the International Neural Network Society for his contributions.

Dr. Principe is a Fellow of the IEEE, Fellow of the AIMBE, past President of the International Neural Network Society, and past Editor in Chief of the Transactions of Biomedical Engineering, as well as a former member of the Advisory Science Board of the FDA. He holds 5 patents and has submitted seven more. Dr. Principe was supervisory committee chair of 50 Ph.D. and 61 Master students, and he is author of more than 400 refereed publications (3 books, 4 edited books, 14 book chapters, 116 journal papers and 276 conference proceedings).

Printed in the United States
by Baker & Taylor Publisher Services